Wellbeing

"From the very first sentence, this timely book wrestles with some of the most pressing issues of our time. An ecological masterpiece, exploring the nuances of our complex relationship with trees, as detailed as the electron microscope imaging used to understand mycorrhizal fungi in the woodland and as far reaching as the broad canopy of a parkland oak. Grounded in culture, history and science, *Wellbeing from Woodland* offers meta-analyses, case studies and human stories, reflecting a depth of love for woodland and commitment to making life better for people and our planet."

"This will be on every Forest School training booklist. It shows why children need woods and woods need children."

—Sarah Lawfull, *FSA Director and Endorsed Trainer*

"When I told a friend I was part of a research project to investigate whether being out in the woods was good for health & wellbeing, he laughed out loud and said "why would you need to research that—everyone knows it's good for you!" And perhaps we all share that gut feeling, an assumption that being out and about in the woods is beneficial for us—it's obvious, isn't it? But is it? Why? And how?"

"The Good from Woods research described in this book goes deeper into these questions: what is it about being in nature that delivers benefits for health and wellbeing? Are woods particularly well suited for it? Is it the place or the activities or the people you're with? Could I get the same benefits just by going for a woodland walk with my dog?"

"For the Forest of Avon Trust, taking part in Good from Woods gave us a chance to really think about what we understood by wellbeing and to take part in live research, helping develop and test a shared framework to record different aspects of wellbeing. It supported us to examine and improve our own practice and focus on making our projects as good as possible in delivering woodland wellbeing to a range of audiences. It is fascinating to read the case studies of the other partner organisations, who worked with different groups and used different research methods, and their findings will prompt more discussion on how to design and develop future projects."

"At a time of so much discussion about the potential benefits for health and wellbeing of being in nature, this is an important book for anyone thinking of commissioning a nature and wellbeing programme, as well as practitioners who are designing or running such woodland projects for health and wellbeing."

—Nicola Ramsden, *Health and Wellbeing Officer, The Forest of Avon Trust*

"The connection between humans and woodlands is old and very rich and this relationship is deeply woven through human cultures within myths and stories. Woodlands have traditionally provided us not only with fuel as firewood, but also refuge and solace in times of trouble. We are now currently living within a time of huge environmental and social uncertainty creating increasing pressure on both natural ecosystems and human health and wellbeing. However, practitioners involved in practical woodland-based activities, such as conservation work or Forest School described in this book, have long recognised woodlands as a rich source of embodied experience that can have a profound and positive impact on human health and wellbeing as well as also benefiting woodlands."

"In this extremely well-researched book, Goodenough and Waite have gathered a wide body of evidence from the literature and diverse collaborative projects to explain how the idea of "woodland wellbeing" can be used to evidence the impact of woodland activities on psychological and wellbeing services."

"This book provides a much-needed evidence base as well as descriptions of best practice and project evaluation that can allow commissioners and service users to understand woodland-based activities and to have full confidence that they will be receiving an effective health and wellbeing service from woodland practitioners within the setting of the woods."

—Roger Duncan, *CAMHS systemic psychotherapist and author of* Nature in Mind, Systemic Thinking and Imagination in Ecopsychology and Mental Health

"The need for people to be able to manage effectively the pressures that daily life brings seems more acute than ever. There is now a very strong body of evidence, which this publication eloquently brings together, which clearly identifies the connection between improved wellbeing and access to greenspaces, particularly woodland. As a forester who has worked amongst trees for over 20 years, I am sure I intrinsically understood this but have only recently, by reading research such as those detailed within, been able to understand what this really means. My hope, along with that of the National Trust is that this information can only help to encourage more people to access the amazing array of greenspaces located across the whole UK and benefit from the increased wellbeing that I am fortunate enough to take for granted."

—John Deakin, *Head of Trees and Woodland at the National Trust, UK*

Alice Goodenough · Sue Waite

Wellbeing from Woodland

A Critical Exploration of Links Between Trees and Human Health

Alice Goodenough
Independent Researcher
Stroud, UK

Sue Waite
Institute of Education
University of Plymouth
Plymouth, UK

ISBN 978-3-030-32631-9 ISBN 978-3-030-32629-6 (eBook)
https://doi.org/10.1007/978-3-030-32629-6

Cover illustration: Maram_shutterstock.com

This Palgrave Macmillan imprint is published by the registered company Springer Nature Switzerland AG
The registered company address is: Gewerbestrasse 11, 6330 Cham, Switzerland

Acknowledgements

We would like to acknowledge and thank BIG Lottery and the University of Plymouth for funding support, Sarah Vaughan and trustees of the Silvanus Trust, and all the practitioner-researchers and their organisations:

Jane Acton and Nature Workshops
Jenny Archard and Otterhead Forest School
Jennie Aronsson and Mayflower Community Academy
Gemma Baal and The Centre for Contemporary Arts and the Natural World
Jade Bartlett and The National Trust
Ali Coles, Seb Buckton, Richard Keating, Jackie Rowanly and Stroud Community Woodland Coop
Katy Lee and Vince Large, Courage Copse Creatives
Victoria Norris, Nicky Puttick and The Woodland Trust
Nicola Ramsden, Rachel Tomlinson, The Forest of Avon Trust and Into the Woods
Richard Turley and Ruskin Mill College

Jessie Watson-Brown and Embercombe
Naomi Wright and Play Torbay

without whom this book could not have been written.

We also owe huge thanks to our families who have been patient and supportive throughout the process because they share similar enthusiasm for trees.

Contents

List of Figures

List of Tables

1

Introduction: The Good from Woods Project

Good from Woods (GfW) was a lottery-funded research project, led by the Silvanus Trust and the University of Plymouth, in partnership with the Forestry Commission, the Neroche Scheme and the Woodland Trust. It ran from April 2010 to December 2014.

It aimed to explore the social and wellbeing outcomes of woodland activities across the south-west of England. The concept of wellbeing entered UK policy on health as part of a movement to reform public health structures and has become intrinsic to modernisation of health services (La Placa & Knight, 2014). Mainstreaming of health and happiness is evident in the reshaping of public health frameworks to reflect local community agendas and a broader range of needs (ibid., p. 6). Restructuring includes the Health and Social Care Act (2012), transferring responsibility for public health and a new duty to promote health to local authorities with the establishment of local Health and Wellbeing Boards (Heath, 2014). The strategic use of wellbeing, associated measures and positive psychology ideas within this devolution is not without its critics (Scott, 2015). Nevertheless, the shift towards tailored localised provision has created new opportunities for different types of wellbeing service delivery to emerge with the third sector, amongst others, supplying health promotion and therapeutic activities conceived around local people and place (La Placa

A. Goodenough and S. Waite, *Wellbeing from Woodland*,
https://doi.org/10.1007/978-3-030-32629-6_1

& Knight, 2014). These activities can be socially prescribed by general practitioners, signposting patients towards projects which can support health and happiness (Bloomfield, 2017).

Initiatives across the south-west deliver a range of woodland-based activities that provide people taking part with personal and social benefits that potentially fit within this devolved and broader conception of health service provision. Activities range from forest education to recreation and involve people of all ages and backgrounds. While each initiative might individually evaluate their success, this information was rarely available for wider use, consideration or demonstration of its benefits. GfW therefore supported organisations to find out and record how people taking part felt about the experience in terms of feeling healthy and happy, worked with providers to develop appropriate research approaches and trained staff to collect evidence, analyse and report it. Tools and findings were shared across the project and with wider communities through workshops, conferences and research articles.

GfW was based on the premise that woodland service providers were well placed to understand, collect and share findings on woodland wellbeing. The project recognised and supported the value of local environmental and social knowledge. Some of the services that engaged with GfW had been developed over a considerable length of time to meet specific needs within wooded settings. Others were in the process of testing newer contexts for the delivery of health and wellbeing in and via the natural world.

Our intention was to

- Contribute to our growing understanding of how outdoor environments benefit the health and wellbeing of individuals and communities;
- Help organisations providing woodland activities understand what makes them effective;
- Help demonstrate to funding bodies the value of supporting woodland activities;
- Potentially increase the opportunities for enjoying the social and wellbeing benefits of activities in woodlands across the south-west.

After the official project's end, GfW has continued as a series of collaborations between its practitioner-researchers and the research team and many of its outputs are stored on the University of Plymouth's Peninsular Research in Outdoor Learning website https://www.plymouth.ac.uk/research/peninsula-research-in-outdoor-learning/good-from-woods.

Why 'Wellbeing from Woodland'?

We each have personal reasons with deep roots anchoring our interest and determination to explore issues around wellbeing from woodland and we would like to share some of these in this introduction to set the context for our book.

Alice: Growing up in a market town, my friends and I spent most of our time wandering out into local countryside to find places where we could relax, explore, take risks and have fun without fear of being told off! We used the local canal as a route in and out of town and I was always excited as we left behind the streets, explored post-industrial spaces being slowly recolonised by nature and emerged into wooded valleys. I loved the deeper immersion into green that these journeys took us on. As I grew up, I watched the small natural spaces within town being filled up by development and expeditions to find somewhere kids could relax and enjoy wilder places become more remote. Later, while at University in London, I had frequent dreams that I could not escape the hard landscaping of the city and was wandering the streets looking for a way out to the green. When taking the train back home, I was often surprised by waves of emotion as we emerged from a last black tunnel into the steep wooded hillsides that I had spent time in when young.

These experiences established my abiding interest in our need for and attachment to green places and what governs and supports access to the wellbeing that I personally found within them. During MA research, I explored the ways in which people invest themselves in different landscapes, experiencing them as an important part of their identity and fiercely defending them from change. My Ph.D. focused on the engagement of young people in the creation/regeneration of green spaces in both urban and rural settings, and the evidence that I found of their attachments to and playful relationship with nature felt familiar. I also listened to young people for whom use of local green environment was uncomfortable (if they were not welcome) and sometimes unsafe (when disruptive or criminal activity took place there). In these circumstances, they were effectively cut off from the valuable effects of nearby nature.

At the same time as researching, I worked as an environmental educationalist and began exploring how supported access to nature-inspired activities helped overcome barriers to young people and adults experiencing natural wellbeing. This led to my interest in Forest Schools and later training as a Forest School leader.

Joining the GfW project in 2010 was an amazing opportunity to pull together my experience as both a researcher and educator and support others to investigate and report on how nature supports our health and happiness. Collaborating with practitioners helped me understand so much about how trees, woods and forests support human flourishing.

Sue: My parents worked in special education and my childhood was spent in residential schools which were often set in magnificent grounds. Although there were few scientific studies evidencing the benefits at that time, locating health and wellbeing promoting institutions in natural surroundings was long established by the 1960s (Hartig et al., 2011). Outside our flat in one school, there was a majestic Cedar of Lebanon where children (including me) would sit cross-legged on the flat rafts of needled branches, like eagles in their nest. A mixed broadleaf wood lay between my home and primary school and my mum and I would walk through the patterns of hazel leaf shadows dappling onto wood anemones, dog's mercury and bluebells, drinking in the air, sounds and smell. Secondary school was also housed in a former manor house and break times were spent playing with pals under rhododendron bushes. At university, I chose a campus with a series of lakes running through woodland and remember the shock to my system when I became a postgraduate and went to live in one of the Pottery towns, prompting me to write a poem/lament 'Not a tree in sight'. Within a term, I had moved to the countryside and trees again. When our first child was born, in early mornings we would cradle her so she could watch the pendulous silver birch leaves outside the window, scintillating green and gold in the sunlight with the refrain 'Oh tree-e- e!' Throughout my life, I have sought out and found huge pleasure in being close to, seeing, listening and touching trees and derived wellbeing from woodland.

Working in the field of outdoor learning at the University of Plymouth, I was involved in a series of projects about Forest School funded by the local early years partnership at Devon County Council. The partnership wanted to explore whether Forest School was something that they should invest money into training. In 2004–2005, Bernie Davis and I supported undergraduate research looking at different aspects of Forest School programmes through which the students also gained experience as assistants to the trained leaders. The students followed their own individual research interests, but the results were aggregated and were mapped to early years learning goals to assess Forest School's usefulness for early years settings (Davis & Waite, 2005; Waite & Davis, 2007).

Following this study, we carried out a survey of Forest School practitioners to tease out the principles underlying practice, a survey of outdoor learning provision across the county and case studies of Forest School in five early years and primary settings: a childminder, a playgroup, a private nursery, a foundation stage and a primary school (Waite, 2015; Waite, Davis, & Brown, 2006). Some issues emerging from these studies are also discussed within this book.

In 2009, the Silvanus Trust and University of Plymouth in partnership with the Woodland Trust and Neroche scheme received a Big Lottery Research Programme award to pursue some of the questions arising from this series of small-scale studies: social cohesion and wellbeing deriving from woodland activities, which we later renamed simply Good from Woods, described above, and this book represents some of the learning that we have been privileged to build over the last decade.

The Organisation of the Book

In Chapter 2, we review literature linking green space and health and then focus on forest studies to explore why woodland per se might support wellbeing. We then consider in Chapter 3 how wellbeing is currently conceived and set out the domains that GfW proposed in their studies across multiple contexts. In the next chapter, we consider some of the challenges to capturing wellbeing outcomes, especially in the mobile contexts of forest activities. Chapters 5–9 then focus on each wellbeing domain in turn and using case studies, discuss how these outcomes are achieved, highlighting the top three messages from the research. In the light of the case studies and other literature, Chapter 10 turns to consider implications for policy and practice in designing means to increase wellbeing through woodland, reflecting on general population green infrastructure and targeted interventions for specific groups. Finally, the main messages from GfW are summarised to argue that we need to encourage a diversity of provision in order to meet varying needs. We share a framework of indicators developed through the project and recommend careful evaluation to help design and refine appropriate programmes.

Over the last decade, we have sent out exploratory shoots in our research, tried to sow seeds in our dissemination and send up suckers of smaller projects and ideas to propagate what we have experienced and evidenced that tree-mendous good can come from woods. We hope that now sharing this work through this book can stimulate similar growth in thinking and practice for others.

References

Bloomfield, D. (2017). What makes nature-based interventions for mental health successful? *British Journal of Psychiatry International, 14*(4), 82–85.

Davis, B., & Waite, S. (2005, January). *Forest schools: An evaluation of the opportunities and challenges in Early Years* (Final Report, report for funding bodies, Devon EYDCP [zero14plus] and the Forest Education Initiative). Plymouth: University of Plymouth.

Hartig, T., van den Berg, A. E., Hagerhall, C. M., Tomalak, M., Bauer, N., Hansmann, R., … Waaseth, G. (2011). Health benefits of nature experience: Psychological, social and cultural processes (Chapter 5). In K. Nilsson, M. Sangster, C. Gallis, T. Hartig, S. de Vries, K. Seeland, & J. Schipperijn (Eds.), *Forests, trees and human health*. New York: Springer.

Health and Social Care Act. (2012). c.7. Retrieved from http://www.legislation.gov.uk/ukpga/2012/7/contents/enacted.

Heath, S. (2014). *Local authorities' public health responsibilities.* London, UK: House of Commons Library.
La Placa, V., & Knight, A. (2014). Well-being: Its influence and local impact on public health. *Public Health, 128*(1), 38–42.
Scott, K. (2015). Happiness on your doorstep: Disputing the boundaries of wellbeing and localism. *The Geographical Journal, 181*(2), 129–137.
Waite, S. (2015). Culture clash and concord: Supporting early learning outdoors in the UK. In H. Prince, K. Henderson, & B. Humberstone (Eds.), *International handbook of outdoor studies.* London: Routledge.
Waite, S., Davis, B., & Brown, K. (2006, July). *Five stories of outdoor learning from settings for 2–11-year olds in Devon* (Final report for funding body EYDCP [zero14plus] and participants). Plymouth: University of Plymouth.
Waite, S., & Davis, B. (2007) The contribution of free play and structured activities in Forest School to learning beyond cognition: An English case. In B. Ravn & N. Kryger (Eds.), *Learning beyond cognition* (pp. 257–274). Copenhagen: The Danish University of Education.

2

Woodland Wellbeing

The 2030 Agenda for Sustainable Development, with its seventeen Sustainable Development Goals (SDGs) (United Nations, 2015), has become the central framework for guiding development policies in countries throughout the world. Sustainability is not just about 'looking after nature' but includes social justice and the wellbeing of all who share this planet through economic, political, social and natural dimensions. There are many studies that have linked natural environments and human wellbeing. In our book, Wellbeing from Woodland, we zoom in on natural and social aspects of woodland experiences in England and how these might support sustainable approaches to wellbeing. In this chapter, we outline the power that green spaces have for human health and wellbeing, before beginning to explore the special place that trees, woods and forest occupy in relation to people, how this varies according to species and cultures, what qualities make woodland special and how certain cultural woodland practices intersect with feelings of wellbeing. In summary, we offer an introduction to woodland wellbeing.

A. Goodenough and S. Waite, *Wellbeing from Woodland*,
https://doi.org/10.1007/978-3-030-32629-6_2

The Power of Green and Our Evolution Within Nature

Our human attachment to trees, woods and forests and capacity to gain wellbeing from spending time in arboreal landscapes can be explored through the lens of our evolution within green environments. Much research about the power of green landscapes concerns human interactions with 'nature' or 'natural environments'. Pinning down exactly what is meant by these terms and specifying effects and relationships is complicated (Wilson, 2019). However, there is strong evidence that green, even on-screen (Lohr & Pearson-Mims, 2006; White et al., 2018) is beneficial for human health and wellbeing. Several theories, summarised below, explain our perceptions of green landscapes and associated impacts on our health and wellbeing as effects of our evolved responses to environment (Hartig et al., 2011). These suggest that human preferences for green are innate responses to landscape that are in no way random but evolved to meet our needs.

Biophilia

The biophilia hypothesis, first proposed by Wilson (1984) and developed by Kellert and Wilson (1995), identifies people's positive responses to nature to be an evolutionary adaptation for enhancing survival. The biophilia concept suggests we possess an innate 'biocentric' affinity with the natural world, that is rooted in our co-evolution (Wilson, 2009). This long-evolved kinship is arguably demonstrated by our emotional reaction towards natural environments and their presence across cultures (Lumber, Richardson, & Sheffield, 2017; Wilson, 2009). Biophobic responses (such as fear of snakes) are seen by some as complementary in heightening chances of human survival through promoting aversion to threats (Ulrich, 1993). There are two subsets of rationale linked to biophilic theory which help explain human orientation towards natural environments:

Savannah theory suggests that we are genetically programmed to prefer landscapes with spreading trees and broad vistas where early humans developed (though Han [2007] points out that evolution in savannah areas is now contested).

Prospect-refuge theory attributes preference for landscapes to survival advantages of seeing without being seen. It is linked closely to functional interpretations of preference where objects are perceived by humans in terms of what they allow, such as Gibson's theory of affordances (1979).

Kahn (1997) questions how far biophilia is genetically determined, whether biophobic reactions contradict the proposed affinity with nature and how the influence of experience and culture may shape any innate propensities.

Environmental Preferences

This theory suggests that the optimally preferred environment is one that has predictability but also elements of complexity and uncertainty that encourage individuals to seek more information. Visual complexity of a landscape, for example, could indicate richness of resources, while patterns and coherence may make a landscape relatively easy to interpret (Kaplan, 1992). Changeable natural environments stimulate preference through their intermediate balance of the new and familiar alongside a heightened state of awareness that is thought to have survival advantages (Kaplan & Kaplan, 1982).

Restoration

Theories about restoration through nature experiences also assume that a person will innately prefer and seek out circumstances they find pleasing and so share ground with the evolutionary-based theories described above. Their distinct contribution is to contrast human responses to natural and built environments and the differing demands these settings place on our cognition. Restoration theories tend therefore to be more psychologically and culturally focused. Two main strands of explanation underpin restoration theorisation:

Psychoevolutionary theory (PET) argues that humans experience involuntary pleasure and consequent recovery from stress in safe natural environments and that this response is the hallmark of an evolved reflex (Ulrich et al., 1991, p. 208). Early humans that recuperated in unthreatening landscapes may have better managed their survival. Our general preference for natural over urban settings (ibid.) may still have evolutionary advantages in managing stress.

Attention restoration theory (ART) suggests our brain processes require rest from directed attention (a focus frequently necessary in modern contexts, but a limited resource), through the comparatively effortless 'soft' fascination of natural environments. The four stages of restoration include: (1) 'clearing the head', (2) recharging directed attention capacity, (3) random unbidden thoughts facilitated by soft fascination and (4) 'reflections on one's life, on one's priorities and possibilities, on one's actions and one's goals' (Kaplan & Kaplan, 1989, p. 197). This affordance of reflection, particularly upon our self and circumstances, may help us resolve distracting issues, further restoring our attention (Basu, Duvall, & Kaplan, 2019).

Taken together, these predominately evolutionary and psychological theories have underpinned many reported associations between green environments and both therapeutic and preventive health and wellbeing outcomes (Hartig et al., 2011).

Testing Hypotheses of Benefits

These hypotheses have inspired numerous experimental studies exploring the association of green with relief of stress and cognitive fatigue in various populations. The siting of trees and plants outside apartment blocks has been argued, for example, to alleviate mental exhaustion and refresh attention amongst female residents (Kuo, 2001; Kuo & Sullivan, 2001), so that women feel better able to manage personal concerns and intrafamily aggression (ibid.). These findings form some of the results from an influential series of studies by Kuo and colleagues in the late 1990s, which led the researchers to propose increasing tree cover as a preventive strategy for avoiding negative social effects. The researchers conducted several large-scale studies in Chicago comparing buildings and spaces with varying levels of tree and grass cover (and controlling for potentially confounding social and environmental factors). Greener spaces were defined by a greater number of trees and their presence transformed barren areas previously viewed with suspicion into spaces well used by local communities and associated with lower levels of graffiti, property and violent crime (Kuo, Bacaicoa, & Sullivan, 1998). Public housing residents also showed a strong preference for images of urban landscapes with more trees, leading Kuo (2003) to conclude that urban forest should be integrated into residential areas.

Focusing on Forest Effects

A number of experimental studies focus on the physiological and psychological aspects of human responses towards woods and forests. A review by Forest Europe (Marušáková & Sallmannshoferet, 2019) found strong evidence that forest visits impact positively on restoration and psychological wellbeing, improve mood and attention and speed psychological stress recovery (Berman, Jonides, & Kaplan, 2008; Berto, Barbiero, Barbiero, & Senes, 2018; Hartig, Evans, Jamner, Davis, & Gärling, 2003; Laumann, Gärling, & Stormark, 2003). It also established there is increasing evidence that time spent in woodland has positive physiological effects, including lowering blood pressure and pulse rate, cortisol levels and sympathetic nervous activity (Martens & Bauer, 2013; Meyer & Kotsch, 2017; Morita et al., 2007; O'Brien & Morris, 2014; Tyrväinen, Lanki, Sipilä, & Komulainen, 2018). For example, female walkers in forests had reduced perceived tension and confusion (Stigsdotter, Sola Corazon, Sidenius, Kristiansen, & Grahn, 2017), while office workers visiting for a day demonstrated lowered blood pressure during their visit and for the following five days (Song, Ikei, & Miyazaki, 2017). It can be difficult, however, to disentangle the influences of the place and the activities undertaken there. Hansmann, Hug, and Seeland (2007) in their study of improvements in stress-related complaints through time spent outdoors found the recovery ratio for stress was 87% on five-point rating scales, but the duration of visits and level of physical activity significantly enhanced positive effects compared to visits involving less strenuous pastimes such as taking a walk or simply relaxing.

Much of the evidence for the impacts of time spent in forests is emerging from Asia, notably Japan, China and Korea, where therapeutic shinrin-yoku or forest walking or 'bathing' inspires research (Kobayashi et al., 2018; Li, 2009; Park, Tsunetsugu, Kasetani, Kagawa, & Miyazaki, 2010). This practice of immersing oneself in nature by mindfully using all five senses is a preventive healthcare initiative that started in the 1980s in Japan. Hansen, Jones, and Tocchini (2017) conducted a systematic review of 64 studies pointing to positive health benefits for the human physiological and psychological systems from this intervention. It has been associated with therapeutic effects on: (1) the immune system function (increase in natural killer cells/cancer prevention); (2) cardiovascular system (hypertension/coronary artery disease); (3) the respiratory system (allergies and respiratory disease); (4) depression and anxiety (mood disorders and stress); (5) mental relaxation (attention deficit hyperactivity disorder); and (6) human feelings of 'awe' (increase in gratitude and selflessness) (Hansen et al., 2017). Within their review, 12 studies specifically addressed stress and stress-related heart disease, emotional distress and chronic depression, alcoholism, sleep disorders and pain. For example, Kim, Lim, Chung, and Woo (2009) found a 4-week forest-walking-based cognitive behavioural therapy programme for treating clinical depression resulted in a significant remission rate of 61% compared to hospital-based therapy (21%). Ideno et al. (2017) compared trials of interventions walking in forest areas (11) or sitting and viewing forest landscapes (7) with the same activities in non-forested city areas (13), sitting in a room (4), or with only measuring blood pressure daily (2). Most trials lasted less than 2 hours (16 trials), but two were longer than 1 day. The meta-analysis showed that the forest environment had a significant effect on lowering blood pressure. Another review of the physiological effects linked to stress reduction of shinrin-yoku (Park, Tsunetsugu, Kasetani, Kagawa, &

Miyazaki, 2009) showed that forest environments could lower blood pressure, sympathetic nerve activity and concentrations of cortisol, and slow pulse rates compared with city settings.

Some of these studies have added a further dimension to stress reduction and attention restoration theorisation, suggesting that the plant life of forests may have a chemical effect on human physiology. Authors exploring improved immune system function and stress reduction following time spent in forests have proposed 'phytoncides', the natural essential oils that trees use to defend themselves from pests, as promoting these outcomes (Lee & Lee, 2014; Li, 2009; Li et al., 2008). Spending time in this unique chemical environment has been proposed both as a form of preventative medicine and immune system support in the treatment of disease such as cancer (Kim, Jeong, Park, & Lee, 2015).

Complicated Associations

These types of findings have been taken up with enthusiasm by audiences seeking to establish how and why green environments support human health and happiness, but there are limitations within the current evidence base. A 2016 systematic review of research exploring ART found, for example, that studies were so diverse in design, definitions, outcomes and participants that comparisons and conclusions were difficult to draw (Ohly et al., 2016). Some key assumptions within ART also remain relatively empirically untested such as the presumed functions of soft fascination (Joye & Dewitte, 2018) with some evidence suggesting it can also characterise attention to natural features not indicative of human survival (Menattia, Subiza-Pérez, Villalpando-Florese, Vozmedianoc, & San Juanc, 2019). Researchers have also cautioned that definition of 'green' space and the metrics measuring exposure to it in experimental studies vary considerably making it difficult to compare effects (Dinand Ekkel & De Vries, 2017). Finally, a number of experimental studies also argue that identity, our subjective sense of who we are, can play an important role within environmental preferences and their capacity to restore us, something that GfW case studies which are explored in subsequent chapters corroborates

(Buijs, Elands, & Langers, 2009; Knez, Ode Sang, Gunnarsson, & Hedblom, 2018; Morton, van der Bles, & Alexander Haslam, 2017; Wilkie & Stavridou, 2013). For example, Lohr and Pearson-Mims (2006) carried out a US-based experiment to see whether emotional and physiological responses to scenes with trees of different forms confirmed the savannah hypothesis. They measured aesthetic preference, affective responses, skin temperature and blood pressure before, during, and after viewing slides of urban scenes with inanimate objects or trees with different forms (comparing responses to a spreading tree or a conical and rounded shaped tree and more open or denser tree canopy). Scenes with trees were all liked more than those with inanimate objects but spreading tree forms and denser canopy were marginally but significantly preferred, which supports explanation of biophilia through savannah and prospect and refuge theories. However, Coss and Moore (2002), in a similar study with 3- to 5-year-old children, found children reported the columnar shape of the familiar Australian pine prettier than an African umbrella thorn tree, but chose the spreading form as preferable to climb, to hide in, to sleep in, or to feel safe from a lion. Such findings suggest that innate responses to nature may be entangled with those that are learnt, in ways not yet fully understood. This evidence reinforces that cultural influence and explanation should warrant careful attention.

Cultural Perspectives on Woodland

Alongside possible innate preference for treed landscapes, our attachment and capacity to feel good in relation to trees, woods and forests varies in relation to social, cultural and economic factors. From a minority world viewpoint, for example, it is easy to forget that non-wood forest products (NWFPs) provide food, income and nutritional diversity for an estimated one in five people around the world, notably women, children, landless farmers and others in vulnerable situations (SOFO, 2018). Trees' material value to humankind varies; for some, an essential living source of food, tools and shelter; for others, felled for timber, fuel and to create space for crops (SOFO, 2018). Human populations are increasing rapidly and expected to reach 10 billion by 2050 with a corresponding 50% increased food demand, putting enormous pressure on forested areas in poor areas to convert to agriculture, but threatening the livelihoods of those dependent on the forest and the continuation of variety of life on Earth. Trees regrow, at least if they are not clear felled. This capacity to withstand many forms of attack and their resilience are perhaps also sources of trees' cultural importance. They not only provide for our physical needs, but inspire endurance and fortitude. As Wilson (2019, p. xvi) suggests, 'Trees do more than nurture our physical bodies; they nurture our spirits and our souls, as well'.

People respond more positively to trees than to other plant types in the landscape (Tahvanainen, Tyrväinen, & Nousianinen, 1996) and they are interwoven into our ideas about who we and our communities are. Many religions, for example, have symbolic trees, like the tree of knowledge in the Garden of Eden and the Bodhi tree beneath which Buddha found enlightenment. In Maori culture, the towering kauri tree is the Son of Land and Sky and God of the Forest. Trees have also been associated with the nurturing of mankind since ancient times. Some species of ash tree exude a sugary substance called *méli*, which was harvested commercially in Greece until recently, and in Norse mythology, Yggdrasil, the world ash, is said to have mead flowing through its branches and to rain honey (http://www.musaios.com/ash.htm). Tree and forest imagery have been central to projects of nation building, for example, the Oak is emblematic of reputed steadfastness and endurance in England (MacNaghten & Urry, 2001; Stafford 2016). While the tree is often valorised in the singular, *en masse* as 'forest', they sometimes elicit more ambivalent views, with many fairy tales depicting woods as dark, unsafe and hostile. Spiritual, literary and cultural meanings enrich the scientific biodiversity represented by the over 60,000 species of trees on our planet. Several studies of public views of woodland in the UK have found that adults perceive trees in relation to such historical and cultural significance, as symbols of nature, life, the state of the environment and British values (Carter, O'Brien, & Morris, 2011; O'Brien, 2005).

There are many woody words beyond forest and woodland, such as copse, spinney, grove, jungle, backwoods, weald, bush, brake, boscage, coppice, chase, plantation, scrub, thicket, greenwood, wildwood, rainforest, which conjure a variety of cultural associations, and this is just in the

English language. England itself only has 13% of wooded land compared to an average of 44% in the rest of Europe and 31% worldwide (SOFO, 2018). Like snow types in Inuit, the density and types of trees within forests within different countries perhaps impact on both the number of associated words and the cultural meanings that they represent. For example, when there was a threat to publicly owned forests in England in 2011, the response was an unanticipated popular outcry. This may have reflected preferences for arboreal landscapes, cultural valuations of this natural heritage, but perhaps also an awareness of the scarcity of this resource is in England—the chance that access might be further restricted hitting a chord with many (BBC, 2011). Plans to privatise large areas of publicly owned forest resulted in 42,000 responses to a public consultation by the Panel appointed to advise on forestry policy in England (Independent Panel on Forestry, 2011). The potential loss of this asset refreshed popular appreciation of woodland, which had been stimulated earlier following widespread felling between 1950 and 1975 and gathered pace during shifts to more mixed woodland planting since the 1990s and the establishment and expansion of the UK's Woodland Trust (Rackham, 2010).

Trees are often seen as totems by long-established communities that had grown up alongside them (Ingold, 1993) carrying meaning across generations, but diverse cultural meanings can collide. Indigenous management of forests in India based on the mutuality of humans and nature were transformed into plundered timber mines during British colonisation (Macnaghten & Urry, 2001). Similar tensions between conflicting values are evident in Levang, Sitorus and Dounias' study (2007), describing how extreme deforestation in Indonesia is entangled within priorities of economic gain and cultural loss that continues to resonate within and across nations. Richard Mabey (2007, p. 9) reflects that:

> The long pattern of our relations with trees begins to look familiar…dependence and notional respect at first; then hubris, rejection, the struggle for dominance and control; then the regret for lost innocence, the return of passion, the pleading for forgiveness.

Some of our capacity to derive wellbeing from spending time in woodlands is thus likely connected to the cultural associations trees and forests hold for us and whether they reinforce a sense of positive personal and collective identity.

Sociodemographic Influences on Woodland Experience

Sociodemographic factors also influence attitudes towards woodland as well as the types of engagement we have with arboreal landscapes (Macnaghten & Urry, 2001; O'Brien, 2004). As MacNaghten and Urry (2001) note in their focus group study of attitudes towards woodland within various social groups (from country sports enthusiasts to inner-city youth), the affordances of treed landscapes, revitalising, relaxing, social, solitary, active or passive, are experienced very differently according to personal and family life-stage, socio-economic circumstance and geographical location. So, for example, ancient broadleaved trees were perceived as symbols of freedom and the English countryside for most groups, but commercial conifer plantations were preferred by Asian young people. This group, who were not interested in spending time amongst trees, was the only one to raise a global ecological need for more woods and forests as a higher priority than personally being in nature. Younger people tended to emphasise use of woodland for being physically active, while outdoor specialist and outdoor enthusiast groups described making use of woods as a rich resource, for foraging flora for food or medicine and developing skills of coppicing, trapping and charcoal making.

A similar focus group study in the south-east and north-west of England by Forest Research (O'Brien, 2004) found woodland was especially valued for social and emotional wellbeing (rather than physical health benefits) through providing an escape from daily pressures. Even in small copses near busy roads, the sense of quiet and being 'away from it all' persisted, although most women regarded visiting forests alone as somewhat risky. Groups tended to attribute feeling good in woods to the multisensory immersion that they provided and to positive cultural, spiritual and historical memories that they stimulated. Their sensory appreciation was at the level of the wood, the individual tree (down to the leaf shapes and tiny lichens growing on them), and the woodland wildlife and flora. In making cultural associations, they referred to Robin Hood, Thomas Hardy and the oak as emblematic of Britain. Generally, people liked safe managed woodland with clear paths and picnic areas and Tyrväinen, Pauleit, Seeland, and de Vries (2005) found that for urbanised people, relative familiarity and perceived safety of natural areas was crucial in enhancing restoration in nature. However, broadleaved woodland was mostly preferred to conifer plantations.

The woodland case study sites explored in subsequent chapters were predominately mixed broadleaf and conscious that the type may influence responses to it, we are careful to describe the character of woodland in each case study. In the light of these variations in response, we would also advocate further research that explores meanings of woodland wellbeing in different countries with different tree species and densities of woodland.

There is interesting evidence that suggests that when natural environments become associated with work their restorative capacity can be limited. Children, for instance, who take part in agriculture in rural areas report reduced restorative experiences compared to children who spend only free time there (Collado, Staats, & Sorrel, 2016). A similar conflict seems to operate for forest workers. Although trees in educational and working environments have been shown to offer wellbeing benefits, forest professionals do not report as much restoration after forest visits as non-forest professionals (von Lindern, Bauer, Frick, Hunziker, & Hartig, 2013), perhaps because of experience fatigue. Nature's restorative qualities differ according to our daily relationship with nature. This might be explained by the concept of cultural density (Waite, 2013).

Cultural Lightness in Woodland

Woodland settings in the UK have fewer cultural associations than their everyday contexts for many people, and novelty is sometimes proposed as supporting engagement with natural environments and allowing new ways of being (Waite & Davis, 2007). Knopf (1987, p. 787) suggests that the unfamiliar culture of green spaces can challenge 'accustomed behaviour patterns, resources, and problem-solving styles'. The non-judgemental indifference of nature permits self-expression, while its relative stability requires less exercise of control. This potential freedom to behave in ways not dictated by our usual habits has also been noted by Waite (2013), who through the concept of institutional habitus suggests that the 'cultural density' of natural environments is often 'light' carrying fewer expected norms of behaviour. Cultural lightness is argued to provide more opportunities to be oneself or try new behaviours, away from the constraints of daily life (ibid.).

Plants as Partners: Interspecies Relational Theories

Other research, drawing on new scientific insights into the 'agency'—capacity to act—of the more-than-human world, has begun to challenge anthropocentric theorisation from evolutionary, psychological and, more recently, cultural perspectives.

Rather than a mere backdrop to human action, plants have been shown to engage in the forms of interactive relations and behaviours, processes which were formerly assumed to be available only to some animals and humans, including perception, memory, learning, decision-making and intra-species communication (Beresford-Kroeger, 2010; Gagliano, 2015, 2017; Gagliano, Mancuso, & Robert, 2012). A blurring of boundaries between human/non-human and what it is be sentient/non-sentient may be necessary to explore people and natural world intra-actions (Stephens, Taket, & Gagliano, 2019). Stephens et al. (2019) argue we need to become more critically aware not only of the dimensions of our personal wellbeing but our responsibilities for the wellbeing of others, including other species (Stephens et al., 2019). This is a world away from the developmental model of nature values proposed by Kellert (2002), which essentially places humankind at the top of Nature's pyramid.

Our current environmental crisis is arguably underpinned by devaluation of plants and microbiota as 'humbler' participants in life on Earth in Western scientific taxonomy (Gagliano, 2013). Gagliano suggests that if we were better able to simply 'be' nature as plants are, we might better appreciate our role within a global network of interrelations. Plants, she says, can:

> teach us to move past the illusion of duality that restrict modern life…and enter a level of entangled reality where there is no time and no separation into self and other, hence no conflict, no destruction, no ecological crisis. Because of this unitive ability to feel at one with life and see the dignity of all manifestations of life, this view of the world cherishes and accepts all beings "as is" in a non-controlling and non-hierarchical way (Gagliano, 2013, p. 6).

Gagliano's suggestion that we take a plant perspective to better understand ourselves and our interrelation with the non-human world can lead to new insights. A tree does not stand-alone, for example, it is a holobiont—dependent on many other organisms and with many organisms depending upon it. Humans can also be regarded as holobionts, assemblages of different species that form ecological units (Mills et al., 2019). Dominant conceptualisations of the relationship between nature and culture are often founded upon an anthropocentric view of what it can do for us. Even recognising that nature and culture merge and we are part of nature; we still face challenges to our comprehension of this as we inevitably see through our human cultural lens. Our very struggle to understand puts us at a disadvantage in fully inhabiting our interconnectedness with nature, as Gagliano (2019) points out.

Nonetheless, nature is demonstrably a network of symbioses with increasing evidence for physical association between individuals of different species for significant parts of their life (Haskell, 2017). This turns the idea of competition being the dominant (or only) impetus on its head, reconfiguring many relationships between species as cooperative and mutual, maintained by communication and a collective intelligence according to some scholars (Gagliano, 2013; Mills et al., 2019; Stephens et al., 2019). The term wood-wide web has been coined to describe the relationship between fungal mycelia, trees and other plants in a complex network of soil ecology (Wohllenberg, 2016). In this broader multi-species picture, communication and learning are not the preserve of 'higher order' mammals. Plants here are sentient—perceiving, feeling, learning, remembering and anticipating—knowing when they are going to be fed or be attacked by pests (e.g. Gagliano & Grimonquez, 2015). Using a biological lens, we can see that plants, for example, interact with their environment through chemicals, conveying messages that are acted upon by recipients. Gagliano and Grimonquez (2015) define language as 'a meaning-making activity at the core of every form of life, including plants' (p. 147). The authors suggest that reconceptualising language as embodied and 'beyond words' acknowledges plants' subjectivity and status and could renew appreciation of our kinship with the more than human. It could be argued that plants and trees along with all matter sharing our planet are our partners—from the Middle English origin of the word as 'joint heir' and 'engaged in the same activity' of being (www.Dictionary.com).

In contrast, others have questioned the need to attribute anthropomorphic intention and purpose to interactions which may be governed more prosaically by genetic programming. Jose, Gillespie, and Pallardy (2004, p. 239), for example, examine a range of possible interspecies relationships, including amensalism, where one species is inhibited and the other unaffected; allelopathy, where one species suppresses the growth or another; commensalism, where one species benefits and the other one is unaffected; competition, where both species are negatively affected as a result of each other's use of resources to grow; mutualism (or synergism), where both species are positively affected; neutralism, where neither species affects the other; and predation and parasitism, where one species consumes the other from the outside or inside. Not all relationships are necessarily positive for all concerned across plants and animals including humans!

However, beneficial networks have been found to exist through holobiont relationships beyond the plant world. Mills et al. (2019) point out that people living in or near more biodiverse environments have less illness and live longer than those from less biodiverse areas, independent of socio-economic status. Such interdependencies suggest human health outcomes can be enhanced by increasing biodiversity with relatively low risk and cost in urban contexts where more and more people are living. Non-human agency, observable in the apparently independent existence of trees, woodland plants, animals and birds, has also been suggested by

Milligan and Bingley (2004) as an appealing quality of woodland for people who are stressed.

Popular Woodland Practices

As we have argued there is a need for more subtlety in understanding what woodland wellbeing comprises. In this next section, some current woodland-based activities in the UK are considered to illustrate contexts within which woodland wellbeing services are being assessed and/or promoted.

The Natural England survey, Monitoring Engagement in Natural Environments (MENE, 2018), has been charting the use of green space in England for a decade. Results from MENE 2015–2016 estimated a total of 446 million visits to woodlands in England during that year; over 100 million more than when the survey began in 2009 (Forest Research, 2018). The MENE surveys also tell us why people visit; 50% of visits were primarily for health and exercise; 38% for walking the dog; and 34% to relax and unwind in 2017/2018 (MENE, 2018).

In a survey by Forest Research (2017) of a representative sample of over 2000 adults aged 16 and over across the UK, 61% of respondents had paid visits to forests or woodlands for walks, picnics or other recreation over the last few years and of those, about three quarters had visited at least once a month during the summer. A much smaller proportion (3%) had been involved actively in woodland conservation, for example, through tree planting, volunteering or community-based woodland. However, more than nine in ten respondents said they saw woodland as a place to relax and de-stress and/or valued them for fun and enjoyment.

Staats and Hartig (2004) found that a forest walk was preferred to a walk in the city when people needed to feel restored and that this benefit was also strongly linked to their expectation that it would do them good. This awareness that some environments are more beneficial may enable people to make better choices for their wellbeing (Korpela, Ylén, Tyrväinen, & Silvennoinen, 2008).

Woodland environments have also long been identified as sites for children's adventurous play. Playlink assessed play space in Forestry Commission owned woodland, suggesting structured play spaces at the entrances to woodland could provide steps towards free play deeper in the forest (Playlink, 2008). This research led to the establishment of The Growing Adventure approach, promoting nature play spaces, environmental play programmes and independent play in woodlands as potential ladders of engagement. The idea is that designed spaces and/or programmed and supervised activities will over time build confidence to explore woodland independently (Gill, 2006). For teenagers too, forest can provide freedom to express themselves. For example, having their own 'territory' to meet up and spend time with friends was valued by young men in Scotland (King, 2010; Morris & O'Brien, 2011; Weldon, Bailey, & O'Brien, 2007).

Another form of woodland practice that is mostly focused on children's engagement and wellbeing is Forest School. What Forest School is remains a contested topic (see, e.g., Leather, 2018; Waite & Goodenough, 2018), but it is gaining momentum across numerous countries worldwide as a means of supporting children's health and wellbeing. Generally, it provides opportunities for children to have regular and frequent contact with woods, to be physically active, to learn through pursuit of their own interests and to become familiar with their local woodlands (Forest School Association, n.d.). There is wealth of research now about Forest School (and examples of practice discussed in later chapters). For example, Roe and Aspinall looked at its role in anger management with young people. Comparing behaviour in school (in a mainstream secondary school and a residential special school), and forest settings and its effects on pupils with 'good' or 'poor' behaviour, they found Forest School helped control anger in young people at risk (Forestry Commission Scotland, 2009).

Studies exploring woodland-based practical conservation activity (such as that organised by the Forestry Commission, Wildlife or Woodland Trusts) have established that effects it can elicit include spiritual connection to nature and its purposes (O'Brien & Morris, 2014), a sense of biophilic wellbeing and nurturing (Waite, Goodenough, Norris, & Puttick, 2016), social connection and heightened physical activity (O'Brien, Townsend, & Ebden, 2008).

Forest bathing or shinrin-yoku has gained recent popularity in the UK with many organisations offering mindful slow walks in woodland. As explored earlier, much of the evidence supporting its delivery as a psychological and physiological woodland-based health services originates in Asia. More research based on Western contexts is needed to begin to explore the effects in different cultures and woodland types.

The National Forest is a huge, new forest plantation intended to regenerate the physical, economic and social landscape across 200 square miles (520 km^2) of central England. Six million trees have increased woodland cover in this region from 6 to 16% in a little over a decade. A study of the impact of the forest has been carried out by Morris and Urry (2006) where evidence was gathered by spending time with people as they engaged in forest- or non-forest-related activities, such as walking, going on-site and educational visits, farm work, tree planting, attending meetings, photography and volunteer work. Sheller and Urry (2006) question conventional distinctions between static 'places' and mobile 'people' who travel to, visit, or access them. They argue that as people move around experiencing but also remembering, imagining and changing, places 'move around' with them. As such, people and place merge in the way that Quay (2017) suggests in his concept of cultureplace. Experience is an amalgam of habitus and the density of cultural norms in places (Waite, 2013), which shapes individual perceptions of and responses to the same events. Thus, for some, tree planting may be seen as the 'new build' in the forest, for others as a memorial for past communities and times of coal mining, while some may reposition farmers as dog wardens to clear up after their dog walking across now open access land. Morris and Urry (2006) highlight how meanings of places are multiple, shifting and contested but that sources of wellbeing can accrue across many different relationships.

The relationship of effects due to different woodland environments, activities and groups of people can be hard to untangle within research reviews and a framework for comparison such as that proposed by Waite, Bølling, and Bentsen (2016) can help to distinguish the factors that are most influential in different contexts. In Chapter 3, we discuss how the template provided for practitioner-researchers scaffolded attention to purposes/aims, content of sessions, the pedagogies used and features of place with specified nuanced outcomes to enable clearer understanding of contributions to woodland wellbeing.

What Is Woodland Wellbeing?

Forest Research (O'Brien & Morris, 2014) synthesised the results of 31 research studies between 2001 and 2012, exploring wellbeing benefits from a wide range of engagement with woodlands in Britain. Personal and public/community benefit survey questions concerned physical wellbeing, nature connectedness, mental wellbeing, education and learning, sense of place, social connectedness and economy. Nature connectedness, mental wellbeing and sense of place were the three most salient wellbeing benefits identified from the meta-analyses.

The first of these is exemplified by a study that looked at transcendent experience in woodland. In 2001, Williams and Harvey in an area of Australia, with dense temperate rainforest, open forests of tall mountain ash, scrubby woodland, eucalypt pockets and pine plantation, conducted a study of 131 people who visited, worked or lived in these forest areas. They asked them to describe a transcendent moment and associated causes, thoughts and behaviours, rating the event for qualities, such as absorption, sense of union and timelessness (ibid.). They found these clustered around three dimensions: fascination, novelty and compatibility.

Fascination was characterised by a feeling of being overwhelmed and fascinated by the forest; belief that the experience was caused by the forest; acute awareness of feelings in body and mind; and description of the environment as complex, full of variety and change.

Novelty was associated with a new experience but included familiarity, arousal and coherence of the environment. Compatibility was linked to feelings of ease, a sense of belonging in the environment and achieving goals and power over the forest. Six types of experience were noted in their study: diminutive, deep flow, non-transcendent, aesthetic, restorative (familiar) and restorative (compatibility), with transcendence encapsulated in diminutive and deep flow experiences. Deep flow experiences were more relaxing and generally attributed to multiple soft foci being present, while diminutive ones were deemed less relaxing and tended to be in response to a strong single focus in the environment. Although 'flow' is often associated with engaging activity (Csikszentmihalyi & Csikszentmihalyi, 1992). Williams and Harvey (2001) found most people attributed deep flow to the material environment itself.

In other work on woodland wellbeing, the presence of other people has been found to be important. Social connections enjoyed during a forest visit were likely to encourage people to spend more time in forests in the future. Three ways in which social connections might support wellbeing have been suggested: through strengthening social relationships; by developing new social relationships; and via participation and community capacity building (Marušáková & Sallmannshoferet, 2019). However, it is often impossible to assess the direction of cause and effect regarding positive feelings from the forest context and/or social benefits. For example, in a large Iranian study of adolescents, it was claimed that more time spent in forests and parks improved self-satisfaction and social contacts (Dadvand et al., 2019). The relationship was associative rather than causal and stronger associations for boys and older adolescents and those in rural areas, as well as those from the lower and higher socio-economic groups were found. As with cultural responses to different types of forest, and complex interrelationships noted between species, nuanced effects are commonly observed.

Milligan and Bingley (2004) similarly note nuance in the ways young people talk about the woods. For example, Tom distinguishes between types of woodland and how they afford different ways of relaxing:

> Some sorts of woodlands seem to lend themselves to particular uses like they're fairly spread out trees so it's quite easy to sort of sit down and whereas some are quite packed together. And it just depends on the woodland, I guess. (Tom, 21 years) (Milligan & Bingley, 2004, pp. 63–64)

Jane, however, found dense forest claustrophobic and liked a view out from the trees, resonating with prospect-refuge theory. Trees themselves were seen as calming and protective, with mature trees imbued with wisdom by some, while other young people emphasised their sensory engagement with details in the colours, scents and texture of woodland for relaxation.

As the Forest Europe report (Marušáková & Sallmannshoferet, 2019) concluded:

> The evidence on the effect of forests on psychological health is not yet good enough to say when, where, and for whom given effects will occur or how long they will last. Positive effects may not be experienced equally by different groups of people (e.g. age, preferences, diseases) and not all types and sizes of forests might be equally effective. (ibid., p. 38)

Thus, it is vital that we take close account of diverse responses and effects for groups and of woodland types in designing interventions as we will discuss further in Chapter 10.

Woodland Wellbeing for Us and for Trees?

Kauppi, Sandstrom, and Lipponen (2018) have traced historical changes in forest cover and Human Development Indicators between 1990 and 2015. Although correlational, not causal links, they suggest that human development and wellbeing can also transform into wellbeing of forest ecosystems, promoting carbon sequestration and global biodiversity. However, this requires a shift from siloed projects for carbon capture, biodiversity conservation or agricultural development to inter-disciplinary and harmonised approaches that balance wellbeing of people and forests (Kauppi et al., 2018). According to Endreny (2018), urban areas occupy 4% of the world's land mass and are expanding, but currently urban forests contain more than 10 billion trees, with over 100 genuses. However, there is further potential in these urban areas for 121 billion trees if planted at global average tree density.

What might be the impacts of the entanglement of nature and culture in woodland wellbeing? In this book, we hope to provoke thinking about woodland and natural/cultural impacts on wellbeing by reporting studies conducted in a particular small and relatively deforested part of the world by forest practitioner-researchers. Stephens et al. (2019) refer to plants as the 'embedded stakeholders of socioecological systems'; many practitioner-researchers referred to trees similarly as they delivered and researched woodland wellbeing impacts. As we turn now to a chapter that frames the complexity of wellbeing, we suggest that the reader carries with them this new ontology and, as we consider the human health and wellbeing outcomes researched as part of the GfW programme, we also reflect on how these activities might impact on the more-than-human world.

To be frank, this was not the position from which we started our programme of work in 2009 but over the years, its sense and centrality has become more and more compelling. Can we achieve sustainability and ecological and social justice through looking only for Good **from** Woods? The cultural service for human wellbeing outcomes has been described as 'nonmaterial benefits people obtain from ecosystems through spiritual enrichment, cognitive development, reflection, recreation, and aesthetic experiences' (Millennium Ecosystem Assessment, 2005a, p. 40, cited in Ambrose-Oji & Fancett, 2011) but as experimental evidence for the cognitive capacities of plants accumulates, and the interweaving of culture and nature is increasingly recognised, an ethics of woodland wellbeing must surely include trees' welfare (Gagliano, 2017).

References

Ambrose-Oji. B., & Fancett, K. (Eds.). (2011). Woods and forests in British Society: Progress in research and practice. In *Conference Proceedings, Forest Research Monograph, 3*. Farnham, Surrey: Forest Research.

Basu, A., Duvall, J., & Kaplan, R. (2019). Attention restoration theory: Exploring the role of soft fascination and mental bandwidth. *Environment and Behavior, 51*(9–10), 1055–1081.

BBC. (2011). *Forest sale axed: Caroline Spelman says, 'I'm sorry'*. Retrieved from http://www.bbc.co.uk/news/uk-politics-12488847.

Beresford-Kroeger, D. (2010). *The Global Forest: 40 ways trees can save us*. London: Penguin Books.

Berman, M. G., Jonides, J., & Kaplan, S. (2008). The cognitive benefits of interacting with nature. *Psychological Science, 19*, 1207–1212.

Berto, R., Barbiero, G., Barbiero, P., & Senes, G. (2018). An individual's connection to nature can affect perceived restorativeness of natural environments. Some observations about Biophilia. *Behavioral Sciences, 8*, 34.

Buijs, A. E., Elands, B. H. M., & Langers, F. (2009). No wilderness for immigrants: Cultural differences in images of nature and landscape preferences. *Landscape and Urban Planning, 91,* 113–123.

Carter, C., O'Brien, L., & Morris, J. (2011). *Enabling positive change: Evaluation of the Neroche landscape partnership scheme.* Farnham: Forest Research.

Collado, S., Staats, H., & Sorrel, M. A. (2016). Helping out on the land: Effects of children's role in agriculture on reported psychological restoration. *Journal of Environmental Psychology, 45,* 201–209.

Coss, R. G., & Moore, M. (2002). Precocious knowledge of trees as antipredator refuge in preschool children: An examination of aesthetics, attributive judgments, and relic sexual dinichism. *Ecological Psychology, 14*(4), 181–222.

Csikszentmihalyi, M., & Csikszentmihalyi, I. S. (Eds.). (1992). *Optimal experience: Psychological studies of flow in consciousness.* Cambridge: Cambridge University Press.

Dadvand, P., Hariri, S., Abbasi, B., Heshmat, R., Qoorbani, M., Motlagh, M. E., … Kelishadi, R. (2019). Use of greenspaces, self-satisfaction and social contacts in adolescents: A population-based CASPIAN-V study. *Environmental Research, 168,* 171–177.

Dinand Ekkel, E., & De Vries, S. (2017). Nearby green space and human health: Evaluating accessibility metrics. *Landscape and Urban Planning, 157,* 214–220.

Endreny, T. A. (2018). Strategically growing the urban forest will improve our world. *Nature Communications. 9* Retrieved from www.nature.com/naturecommunications.

Forestry Commission Scotland. (2009). *Forest School: Evidence for restorative health benefits in young people.* Retrieved from http://www.openspace.eca.ed.ac.uk/wp-content/uploads/2015/10/Forest-school-evidence-for-restorative-health-benefits-in-young-people.pdf.

Forest Research. (2017). *Public opinion of forestry 2017.* Retrieved from http://www.forestry.gov.uk/forestry/infd-5zyl9w.

Forest Research. (2018). *Visits to woodland—Household surveys.* Retrieved from https://www.forestresearch.gov.uk/tools-and-resources/statistics/forestry-statistics/forestry-statistics-2018/recreation/visits-to-woodland-household-surveys/.

Forest School Association. (n.d.). *What is Forest School.* Retrieved from https://www.forestschoolassociation.org/what-is-forest-school/.

Gibson, J. (1979). *The ecological approach to visual perception.* Boston: Houghton Mifflin.

Gagliano, M. (2013). Persons as plants: Ecopsychology and the return to the dream of nature. *Landscapes: The Journal of the International Centre for Landscape and Language, 5*(2). Retrieved from http://ro.ecu.edu.au/landscapes/vol5/iss2/14.

Gagliano, M. (2015). In a green frame of mind: Perspectives on the behavioural ecology and cognitive nature of plants. *AoB Plants, 7,* 1–8.

Gagliano, M. (2017). The mind of plants: Thinking the unthinkable. *Communicative & Integrative Biology, 10*(2), 38427.

Gagliano, M. (2019). *Can plants think, talk and heal?* Retrieved from https://www.youtube.com/watch?v=UF2JMRAt-eI.

Gagliano, M., & Grimonprez, M. (2015). Breaking the silence—Language and the making of meaning in plants. *Ecopsychology, 7*(3), 145–152. https://doi.org/10.1089/eco.2015.0023.

Gagliano, M., Mancuso, S., & Robert, D. (2012). Towards understanding plant bioacoustics. *Trends in Plant Science, 17,* 323–325.

Gill, T. (2006). *Growing adventure.* Retrieved from http://www.forestschoolwales.org.uk/wp-content/uploads/forestry-commission-report-growing-adventure.pdf.

Han, K. T. (2007). Responses to six major terrestrial biomes in terms of scenic beauty, preference, and restorativeness. *Environmental Behavior, 39,* 529–556.

Hansen, M. M., Jones, R., & Tocchini, K. (2017). Shinrin-yoku (forest bathing) and nature therapy: A state-of-the-art review. *International Journal of Environmental Research and Public Health, 14,* 851.

Hansmann, R., Hug, S.-M., & Seeland, K. (2007). Restoration and stress relief through physical activities in forests and parks. *Urban Forestry & Urban Greening, 6*(4), 213–225.

Hartig, T., Evans, G. W., Jamner, L. D., Davis, D. S., & Gärling, T. (2003). Tracking restoration in natural and urban field settings. *Journal of Environmental Psychology, 23,* 109–123.

Hartig, T., van den Berg, A. E., Hagerhall, C. M., Tomalak, M., Bauer, N., Hansmann, R., … Waaseth, G. (2011). Health benefits of nature experience: Psychological, social and cultural processes (Chapter 5). In K. Nilsson, M. Sangster, C. Gallis, T. Hartig, S. de Vries, K. Seeland, & J. Schipperijn (Eds.), *Forests, trees and human health.* New York: Springer.

Haskell, D. G. (2017). *The songs of trees: Stories from nature's great connectors.* New York: Viking.

Ideno, Y., Hayashi, K., Abe, Y., Ueda, K., Iso, H., Noda, M., … Suzuki, S. (2017). Blood pressure-lowering effect of Shinrin-yoku (forest bathing): A systematic review and meta-analysis. *BMC Complementary and Alternative Medicine, 17*(1), 409.

Independent Panel on Forestry. (2011). *Progress report.* London: DEFRA. Retrieved from https://webarchive.nationalarchives.gov.uk/20131001175003/, http://www.defra.gov.uk/forestrypanel/files/Independent-Panel-on-Forestry-Progress-Report.pdf.

Ingold, T. (1993). The temporality of the landscape. *World Archaeology, 25*(2), 152–174.

Jose, S., Gillespie, A. R., & Pallardy, S. G. (2004). Interspecific interactions in temperate agroforestry. *Agroforestry Systems, 61,* 237–255.

Joye, Y., & Dewitte, S. (2018). Nature's broken path to restoration. A critical look at attention restoration theory. *Journal of Environmental Psychology, 59,* 1–8.

Kahn, P. H. (1997). Developmental psychology and the biophilia hypothesis: Children's affiliation with nature. *Development Review, 17,* 1–61.

Kaplan, S., & Kaplan, R. (1982). *Cognition and environment: Functioning in an uncertain world.* New York: Praeger.

Kaplan, S., & Kaplan, R. (1989). *The experience of nature: A psychological perspective.* Cambridge: Cambridge University Press.

Kaplan, S. (1992). Perceptions and landscape: Conceptions and misconceptions. In J. Nasar (Ed.), *Environmental aesthetics: Theory, research, and application.* New York: Cambridge University Press.

Kauppi, P. E., Sandstrom, V., & Lipponen, A. (2018). *Forest resources of nations in relation to human well-being. PLoS ONE, 13*(5). Retrieved from https://doi.org/10.1371/journal.pone.0196248.

Kellert, S., & Wilson, E. O. (1995). *The Biophilia hypothesis.* Washington, DC: Island Press.

Kellert, S. R. (2002). Experiencing nature: Affective, cognitive, and evaluative development in children. In P. H. Kahn & S. R. Kellert (Eds.), *Children and nature: Psychological, sociocultural, and evolutionary investigations* (pp. 117–152). Cambridge: MIT Press.

Kim, B. J., Jeong, H., Park, S., & Lee, S. (2015). Forest adjuvant anti-cancer therapy to enhance natural cytotoxicity in urban women with breast cancer: A preliminary prospective interventional study. *European Journal of Integrative Medicine, 7*(5), 474–478.

Kim, W., Lim, S., Chung, E., & Woo, J. (2009). The effect of cognitive behavior therapy-based psychotherapy applied in a forest environment on physiological changes and remission of major depressive disorder. *Psychiatry Investigations, 6,* 245–254.

King, K. (2010). *Lifestyle, identity and young people's experiences of mountain biking.* Edinburgh: Forestry Commission.

Knez, I., Ode, Sang Å., Gunnarsson, B., & Hedblom, M. (2018). Wellbeing in urban greenery: The role of naturalness and place identity. *Frontiers in Psychology, 9,* 491.

Knopf, R. (1987). Human behavior, cognition, and affect in the natural environment. In D. Stokols & I. Altman (Eds.), *Handbook of environmental psychology* (Vol. 1, pp. 783–825). New York: Wiley.

Kobayashi, H., Song, C., Ikei, H., Park, B. J., Lee, J., Kagawa, T., & Miyazaki, Y. (2018). Forest walking affects autonomic nervous activity: A population-based study. *Public Health, 6,* 278.

Korpela, K. M., Ylén, M., Tyrväinen, L., & Silvennoinen, H. (2008). Determinants of restorative experiences in everyday favorite places. *Health & Place, 14*(4), 636–652.

Kuo, F. E. (2001). Coping with poverty impacts of environment and attention in the inner city. *Environment and Behavior, 33*(1), 5–34.

Kuo, F. E., Bacaicoa, M., & Sullivan, W. C. (1998). Transforming inner-city landscapes: Trees, sense of safety, and preference. *Environment and Behavior, 30,* 28–59.

Kuo, F. E., & Sullivan, W. C. (2001). Aggression and violence in the inner-city effects of environment via mental fatigue. *Environment and Behavior, 33*(4), 543–571.

Kuo, F. E. (2003). The role of arboriculture in a healthy social ecology. *Journal of Arboriculture, 29*(3), 148–155.

Laumann, K., Gärling, T., & Stormark, K. M. (2003). Selective attention and heart rate responses to natural and urban environments. *Journal of Environmental Psychology, 23*(2), 125–134.

Leather, M. (2018). A critique of Forest School: Something lost in translation. *Journal of Outdoor and Environmental Education., 21*(1), 5–18.

Lee, J. Y., & Lee, D. C. (2014). Cardiac and pulmonary benefits of forest walking versus city walking in elderly women: A randomised, controlled, open-label trial. *European Journal of Integrative Medicine., 6*(1), 5–11.

Levang, P., Sitorus, S., & Dounias, E. (2007). City life in the midst of the forest: A Punan hunter-gatherer's vision of conservation and development. *Ecology and Society, 12*(1), 18. Retrieved from https://www.ecologyandsociety.org/vol12/iss1/art18/.

Li, Q. (2009). Effect of forest bathing trips on human immune function. *Environmental Health and Preventive Medicine, 15*(1), 9–17.

Li, Q., Morimoto, K., Kobayashi, M., Inagaki, H., Katsumata, M., Hirata, Y., … Krensky, A. M. (2008). Visiting a forest, but not a city, increases human natural killer activity and expression of anti-cancer proteins. *International Journal of Immunopathology and Pharmacology,* 21, 117–127.

Lohr, V. I., & Pearson-Mims, C. H. (2006). Responses to scenes with spreading, rounded and conical tree forms. *Environment and Behavior, 38,* 667–688.

Lumber, R., Richardson, M., & Sheffield, D. (2017). Beyond knowing nature: Contact, emotion, compassion, meaning, and beauty are pathways to nature connection. *PLOS ONE, 12*(5). Retrieved from https://journals.plos.org/plosone/article?id=10.1371/journal.pone.0177186#sec001.

Mabey, R. (2007). *Beechcombings: The narratives of trees.* London: Chatto & Windus.

Macnaghten, P., & Urry, J. (2001). Bodies in the woods. In P. Macnaghten & J. Urry (Eds.), *Bodies of nature.* London: Sage.

Martens, D., & Bauer, N. (2013). Natural environments—A resource for public health and wellbeing: A literature review. In E. Noethammer (Ed.), *Psychology of wellbeing: Theory, perspectives and practice* (pp. 173–217). New York: Nova Science Publishers.

Marušáková, Ľ., & Sallmannshoferet, M. (2019). *Human health and sustainable forest management.* Bratislava: FOREST EUROPE. Retrieved from https://foresteurope.org/wp-content/uploads/2017/08/Forest_book_final_WEBpdf.pdf.

Menattia, L., Subiza-Pérez, M., Villalpando-Florese, A., Vozmedianoc, L., & San Juanc, C. (2019). Place attachment and identification as predictors of expected landscape restorativeness. *Journal of Environmental Psychology, 63,* 36–43.

MENE. (2018). *Monitoring engagement in the natural environment.* Retrieved from https://assets.publishing.service.gov.uk/government/uploads/system/uploads/attachment_data/file/738891/Monitorof_Engagementwiththe_Natural_Environment_Headline_Report_March_2016to_February_2018.pdf.

Meyer, K., & Kotsch, K. (2017). Do forest and health professionals presume that forests offer health benefits, and is cross-sectional cooperation conceivable? *Urban Forestry & Urban Greening, 27,* 127–137.

Milligan, C., & Bingley, A. (2004). *'Climbing trees and building dens': Mental health and well-being in young adults and the long-term effects of childhood play experience.* Farnham: Forest Research.

Mills, J. G., Brookes, J. D., Gellie, N. J. C., Liddicoat, C., Lowe, A. J., Sydnor, H. R., … Breed, M. F. (2019). Relating urban biodiversity to human health with the 'holobiont' concept. *Frontiers in Microbiology, 10*, 550.

Morita, E., Fukuda, S., Nagano, J., Hamajima, N., Yamamoto, H., Iwai, Y., … Shirakawa, T. (2007). Psychological effects of forest environments on healthy adults: Shirinyoku (forest air bathing, walking) as a possible method of stress reduction. *Public Health, 121*(1), 54–63.

Morton, T. A., van der Bles, A. M., & Alexander Haslam, S. (2017). Seeing our self reflected in the world around us: The role of identity in making (natural) environments restorative. *Journal of Environmental Psychology, 49*, 65–77.

Morris, J., & Urry, J. (2006). *Growing places: A study of social change in The National Forest.* Farnham: Forest Research.

Morris, J., & O'Brien, L. (2011). Encouraging healthy activity amongst under-represented groups: An evaluation of the active England woodland projects. *Urban Forestry and Urban Greening, 10*, 323–333.

O'Brien, L. (2004). *A sort of magical place: People's experiences of woodlands in northwest and southeast England.* Farnham: Forest Research.

O'Brien, L. (2005). Publics and woodlands: Well-being, local identity, social learning, conflict and management. *Forestry, 78*, 321–336. https://doi.org/10.1093/forestry/cpi042.

O'Brien, L., & Morris, J. (2014). Well-being for all? The social distribution of benefits gained from woodlands and forests in Britain. *Local Environment, 19*(4), 356–383.

O'Brien, L., Townsend, M., & Ebden, M. (2008). *'I like to think when I'm gone, I will have left this a better place': Environmental volunteering: Motivations, barriers and benefits.* Farnham: Forest Research.

Ohly, H., White, M. P., Wheeler, B. W., Bethel, A., Ukoumunne, O. C., Nikolaou, V., & Garside, R. (2016). Attention restoration theory: A systematic review of the attention restoration potential of exposure to natural environments. *Journal of Toxicology and Environmental Health, Part B, 19*(7), 305–343.

Park, B. J., Tsunetsugu, Y., Kasetani, T., Kagawa, T., & Miyazaki, Y. (2009). The physiological effects of Shinrin-yoku (taking in the forest atmosphere or forest bathing): Evidence from field experiments in 24 forests across Japan. *Environmental Health and Preventive Medicine, 15*, 18–26.

Park, B. J., Tsunetsugu, Y., Kasetani, T., Kagawa, T., & Miyazaki, Y. (2010). The physiological effects of Shinrin-yoku (taking in the forest atmosphere or forest bathing): Evidence from field experiments in 24 forests across Japan. *Environmental Health and Preventive Medicine, 15*, 8–26.

Playlink. (2008). *Forestry commission assessment of provision for play*. Retrieved from https://playlink.org/pdf/fcfinal.pdf.

Quay, J. (2017). From human–nature to cultureplace in education via an exploration of unity and relation in the work of Peirce and Dewey. *Studies in Philosophy and Education, 36*(4), 463–476.

Rackham, O. (2010). *Woodlands*. London: HarperCollins.

Sheller, M., & Urry, J. (2006). The new mobilities paradigm. *Environment and Planning, 38*, 207–226.

SOFO. (2018). *The state of the world's forest*. Retrieved from http://www.fao.org/state-of-forests/en/.

Song, C., Ikei, H., & Miyazaki, Y. (2017). Sustained effects of a forest therapy program on the blood pressure of office workers. *Urban Forestry & Urban Greening., 27*, 246–252.

Staats, H., & Hartig, T. (2004). Alone or with a friend: A social context for psychological restoration and environmental preferences. *Journal of Environmental Psychology, 24*, 199–211.

Stafford, F. (2016). *The long, long life of trees*. London: Yale University Press.

Stigsdotter, U. K., Sola Corazon, S., Sidenius, U., Kristiansen, J., & Grahn, P. (2017). It is not all bad for the grey city—A crossover study on physiological and psychological restoration in a forest and an urban environment. *Health & Place, 46*, 145–154.

Stephens, A., Taket, A., & Gagliano, M. (2019). Ecological justice for nature in critical systems thinking. *Systems Research and Behavioral Science, 36*, 3–19.

Tahvanainen, L., Tyrväinen, L., & Nousianinen, I. (1996). Effects of afforestation on the scenic value of rural landscape. *Scandinavian Journal of Forest Research, 11*, 397–405.

Tyrväinen, L., Pauleit, S., Seeland, K., & de Vries, S. (2005). Benefits and uses of urban forests and trees. In K. Nilsson, T. B. Randrup, & C. C. Konijnendijk (Eds.), *Urban forests and trees in Europe: A reference book* (pp. 81–114). Switzerland: Springer.

Tyrväinen, L., Lanki, T., Sipilä, R., & Komulainen, J. (2018). What do we know about health benefits of forests. *Duodecim, 134*(13), 1397–1403. Retrieved from https://www.duodecimlehti.fi/lehti/2018/13/duo14421.

Ulrich, R. S. (1993). Biophilia, Biophobia and natural landscapes'. In S. R. Kellert & E. O. Wilson (Eds.), *The Biophilia hypothesis*. Washington, DC: Island Press.

Ulrich, R. S., Simons, R. F., Losito, B. D., Fiorito, E., Miles, M. A., & Zelson, M. (1991). Stress recovery during exposure to natural and urban environments. *Journal of Environmental Psychology, 11*(3), 201–230.

United Nations. (2015). *The sustainable development goals*. Retrieved from https://www.un.org/sustainabledevelopment/.
von Lindern, E., Bauer, N., Frick, J., Hunziker, M., & Hartig, T. (2013). Occupational engagement as a constraint on restoration during leisure time in forest settings. *Landscape and Urban Planning, 118*, 90–97.
Waite, S. (2013). 'Knowing your place in the world': How place and culture support and obstruct educational aims. *Cambridge Journal of Education, 43*(4), 413–434.
Waite, S., & Davis, B. (2007). The contribution of free play and structured activities in Forest School to learning beyond cognition: An English case. In B. Ravn & N. Kryger (Eds.), *Learning beyond cognition* (pp. 257–274). Copenhagen: The Danish University of Education.
Waite, S., Bølling, M., & Bentsen, P. (2016). Comparing apples and pears? A conceptual framework for understanding forms of outdoor learning through comparison of English Forest Schools and Danish *udeskole*. *Environmental Education Research, 22*(6), 868–892.
Waite, S., & Goodenough, A. (2018). What is different about Forest School? *Journal of Outdoor and Environmental Education, 21*(1), 25–44.
Waite, S., Goodenough, A., Norris, V., & Puttick, N. (2016). From little acorns: Environmental action as a source of ecological wellbeing. *Pastoral Care in Education: An International Journal of Personal, Social and Emotional Development, 34*(1), 43–61.
Weldon, S., Bailey, C., & O'Brien, E. (2007). *New pathways to health and well-being in Scotland: Research to understand and overcome barriers to accessing woodlands*. Report to Forestry Commission Scotland. Retrieved from https://www.forestresearch.gov.uk/documents/.../New_Pathways_to_health_Nov2007.
White, M., Yeo, N. L., Vassiljev, P., Lundstet, R., Wallegard, M., Albin, M., & Lohmus, M. (2018). A prescription for "nature"—The potential of using virtual nature for therapeutics. *Neuropsychiatric Disease and Treatment, 12*, 3001–3013.
Wilkie, S., & Stavridou, A. (2013). Influence of environmental preference and environment type congruence on judgments of restoration potential. *Urban Forestry & Urban Greening, 12*(2), 163–170.
Williams, K., & Harvey, D. (2001). Transcendent experience in forest environments. *Journal of Environmental Psychology, 21*, 249–260.
Wilson, O. (1984). *Biophilia*. Cambridge, MA: Harvard University Press.

Wilson, E. O. (2009). Biophilia and the conservation ethic. In D. J. Penn & I. Mysterud (Eds.), *Evolutionary perspectives on environmental problems* (pp. 249–258). New Brunswick and London: Aldine Transaction.
Wilson, R. (2019). *Trees and the human spirit.* Newcastle-upon-Tyne: Cambridge Scholars Publishing.
Wohllenberg, P. (2016). *The hidden life of trees: What they feel, how they communicate—Discoveries from a secret world.* London: Greystone Books.

3

Framing Complexity in Wellbeing

In this chapter, we establish the UK context for using wellbeing as a measure of health and happiness in woodland service delivery and explore what types of wellbeing assessment might be relevant in a woodland context and the value of subjective measurements in this environment. We detail what woodland wellbeing might be expected to consist of and introduce the emotional, psychological, social, physical and biophilic domains and indicators of woodland wellbeing employed within the GfW framework for practitioner-researchers.

Wellbeing and Green Space Impetus in UK

At the time of the GfW project's development and launch (2010), the accomplishment and measuring of our nation's wellbeing had become a key concern of UK governance, service provision, NGO campaigns and research (Abdallah, Steuer, Marks, & Page, 2008; Bacon, Brophy, Mguni, Mulgan, & Shandro, 2010; O'Brien, 2009; Thompson & Marks, 2008). Recent recession had focused attention on how the best returns on investment in national happiness could be achieved: establishing, protecting and sustaining it. David Cameron, Prime Minister at the time, suggested that

A. Goodenough and S. Waite, *Wellbeing from Woodland*,
https://doi.org/10.1007/978-3-030-32629-6_3

gross domestic product (GDP) was an 'incomplete' measure of progress and announced that the National Office of Statistics was to establish a 'wellbeing index' that could track happiness and health, in addition to economic growth (Mulholland & Watt, 2010).

Within this context, interest in establishing and measuring how time spent in natural, outdoor spaces contributes to our wellbeing was similarly burgeoning. Activities in green spaces were increasingly being championed by academics, the third sector and commissioners as tools for improving health and happiness. The idea that a lack of nature could in fact negatively impact wellbeing was also gaining strength (Louv, 2005).

During GfW's lifetime and subsequently, a continued focus on and shift in how we understand our happiness in the UK have moved positive psychology perspectives on wellbeing into the mainstream (Horowitz, 2018). The notion of wellbeing has become omnipresent, repositioning our ideas around poor health and its treatment through emphasising preventative approaches to maintaining positive functioning and fitness (White, 2018). The natural world is now viewed as having a key role in this supporting and sustaining of wellbeing: a natural health service.

Why Wellbeing as a Measure of What's Good from Woods?

In the instance of GfW, 'wellbeing' was embraced as an idea and language that could support development of a common approach to unearthing and explaining the outcomes of woodland activity.

A look at conceptions of wellbeing within policy and practice demonstrates its wide interpretation and application. Individual wellbeing can be interpreted, for example, as 'happiness', 'quality of life', 'life satisfaction', 'welfare' or the appropriate personal 'resources' to balance adversity (Bacon et al., 2010, p. 8; Dodge, Daly, Huyton, & Sanders, 2012; O'Brien, 2009, p. 1; Thompson & Marks, 2008, p. 8). It can describe positive physical, social, psychological and emotional attitudes as well as material health and assets (ibid.). This breadth and flexibility of concept was anticipated to allow development of a holistic portrait of the wide range of services provided by woodland activity. It also meant that the project did not have to impose restrictions upon the areas of health and happiness that practitioners might research. At the same time, with the idea of wellbeing at the heart of the project, findings might find some comparison and correspondence with other forms of delivery and research in the field of health and happiness.

While the breadth of the wellbeing concept has distinct advantages for understanding woodland wellbeing, it also requires some sense of shared interpretation to be useful. While GfW supported an action research approach to investigating wellbeing in woodland activity contexts, practitioner-researchers were keen to be directed in understanding what wellbeing, during or following woodland activity, might be composed of. What might it feel or look like? Would it differ in woodland from elsewhere and how should they capture evidence of its presence? Developing a sense of what woodland wellbeing might be is essential to establishing ways of capturing its presence and measuring its significance.

Woodland wellbeing could be defined and measured, for example, through deciding which human needs might be met in that environment that would help us to thrive (physiological, psychological, pedagogical, etc.) and measuring corresponding successes and shortfalls in these areas: an '*objective-list account*' of wellbeing (Dolan & White, 2007, p. 71). It might also be described and calculated through capturing how woodland activity contributes to participants' financial status—their economic ability to satisfy their needs and preferences: a '*desire-fulfilment account*' of wellbeing (Dolan & White, 2007, p. 71; O'Brien, 2009). Its description and achievement could even be mapped through measuring the physiological processes that occur within them (Bacon et al., 2010). Equally, it might be characterised and gauged through exploring people's own subjective assessments of their health and happiness: a '*mental-state account*' of wellbeing (Bacon et al., 2010; Dolan & White, 2007, p. 71).

Defining and appraising wellbeing is not only complicated by these varying definitions (psychological, psychological, societal and financial), but also by differences in the ingredients and intensity of its effects (Bacon et al., 2010).

Understanding, Defining and Measuring Wellbeing

The popularity and use of particular definitions and approaches to measuring wellbeing can vary in relation to the ideas and motivations of those undertaking the evaluation (Dolan & White, 2007). Objective list measurements of wellbeing have been widely deployed within governance, for example. However, these lists cataloguing the conditions necessary for health and happiness embody the values of their designers, reflecting culturally specific ideas of wellbeing (Dolan & White, 2007). Further, one aspect of objective wellbeing can also conflict with another: taxation to provide health and education welfare clashing with the ability of people to achieve the benefits associated with homeownership, for example.

Economic calculations of physical, psychological and social satisfaction have also been identified as sometimes insensitive indicators of wellbeing, with the 'Easterlin paradox' a frequently cited demonstration of this. This phenomenon concerns the contradiction that rising levels of GDP across much of the Western world have not resulted in parallel levels of increased happiness (Adler & Seligman, 2016; Dolan & White, 2007; Jordan, 2008; Michaelson, Abdallah, Steuer, Thompson, & Marks, 2009). Some argue that in fact, fiscal measurement of wellbeing can obscure the negative influences of economic development upon our ability to thrive (from long working days to environmental degradation) and foster policies that ignore or perpetuate these contributors to poor health and life satisfaction (Dolan & White, 2007; Jordan, 2008; Michaelson et al., 2009).

During the lifetime of GfW, there was a growing call for progress beyond the material and economic measures of wellbeing that had previously been the mainstay of gauging our populace's welfare (Allin & Hand, 2017). Instead, there was a shift towards advocating for and adopting evaluations that capture people's personal assessments of their health and happiness: their subjective wellbeing (SWB) (Dolan, Layard, & Metcalfe, 2011; Dolan & White, 2007; Michaelson et al., 2009).

Subjective Wellbeing (SWB)—Life Satisfaction, Emotions and Moods

SWB is a self-directed intuitive approach to describing and determining personal happiness and health (Diener, 2000; Dolan & White, 2007). It is the method we spontaneously employ to estimate our levels of contentment and how we are getting on. Furthermore, a person's self-assessment of their wellbeing has been shown to correlate well with how observers think they are feeling and even the physiological processes occurring within them (Dolan & White, 2007; Helliwell & Putnam, 2004).

SWB can be measured through recording levels of self-reported satisfaction and feelings in three areas of awareness (Diener, 2000, p. 34):

> 'life satisfaction' (a broad-spectrum judgement of overall satisfaction or happiness in significant 'domains', like work, for example),
> 'positive affect' (experiences of agreeable frames of mind and feelings),
> 'negative affect' (disagreeable dispositions and feelings).

Life satisfaction is the cognitive element of SWB or our intellectual assessment of how we are doing in life generally or in specific areas of it. Positive and negative affect is the emotional dimension of SWB or our experiences of positive or negative feelings, moods and emotions. Each area can be independent of the others, so we may feel we are doing well in life but still experience negative moods.

Measuring SWB and the success of interventions aimed at its improvement may be complicated by factors that could cause changes within it to be relatively fleeting. It has been proposed that human behaviours, including 'adaption', 'attention' and 'relative judgement', will tend to return wellbeing to a baseline state in the long term, regardless of temporary peaks and troughs (Bacon et al., 2010; Brekke & Howarth, 2002; Diener, Lucas, & Napa Scollon, 2006; Dolan & White, 2007; Frank, 1999).

Adaption describes people's tendency to become accustomed to changed living conditions, physical, social, financial, etc. In this way, difficult life circumstances can become by subjective measures, customary and acceptable, while remaining by objective measures, unsatisfactory. So, for example, people living in disadvantaged conditions may report feelings and satisfactions similar to those in evidently less difficult situations (Dolan & White, 2007).

The consistency of SWB accounts may also be affected by the way in which individuals chose to focus attention upon particular areas of their life, rather than others. It may not be adaptation that triggers relatively high perceptions of wellbeing amongst those objectively facing adverse circumstances, but the choice people make to pay attention to areas of their lives that are more satisfactory (Dolan & White, 2007; Kahneman & Krueger, 2006).

A human inclination to adapt to life circumstances may also reduce the felt impact of objective improvements within them, trapping us on a 'hedonic treadmill' (ibid.). Incremental gains in satisfaction can be adapted to, become customary and lose their significance. This can shift our attention towards new aspirations for health and happiness and efforts to achieve them, leaving us feeling that we are making little progress in our goals (Brekke & Howarth, 2002; Dolan & White, 2007). This treadmill effect may be one driver of the Easterlin paradox: that satisfaction is a brief state and therefore aspiration for satisfaction is perpetual.

Self-assessment of wellbeing in prosperous societies appears to be partially governed by our estimations of how we compare with others—relative judgements of our financial, social and physical status (Brekke & Howarth, 2002; Frank, 1999; Helliwell & Putnam, 2004). This comparative consideration of how we are doing in contrast to other people helps determine aspirations for happiness and good health. These relative judgements appear to fuel aspirations for 'positional goods', attainments that are relatively rare and suggest superior social status such as high levels of income, education and physical attractiveness (Bacon et al., 2010, p. 25; Brekke & Howarth, 2002, pp. 40–43). These cultural and social influences upon our sense of wellbeing may be additional drivers of the hedonic treadmill and the Easterlin paradox (Adler & Seligman, 2016; Bacon et al., 2010; Brekke & Howarth, 2002; Helliwell & Putnam, 2004). Easterlin has more recently revised his model of wealth's relation to happiness to suggest that perceptions of relative wealth are what are fundamental to life satisfaction (Adler & Seligman, 2016). If there is income inequality within countries, or between nations, then despite absolute wealth gains populations will remain dissatisfied (ibid.). As sociocultural influences redefine what is believed necessary for welfare and social status, they persistently shift and fuel aspirations for standards of wellbeing: a recipe for happiness with an ever-changing (and scarce) list of ingredients.

However, the idea that SWB will eventually return to a set level or point, despite the impact of events, has been challenged by research which suggests these points change and are variable. Levels of SWB do appear in fact to respond to who we are: our genetic inheritance, cultural background and the point we have reached in our life course (Diener et al., 2006; Dolan & White, 2007). For instance, there is evidence to suggest that people have different dispositions, personality traits that may result in them feeling generally more positive or negative than someone else (Diener, 2000; Diener et al., 2006). Such traits might be part of our genetic inheritance or shaped by our early life experiences and cultural context and may mean that people have different base levels of SWB. The rate and extent to which people adapt to alterations in levels of wellbeing can also vary between individuals, some people tending to recover from difficult events much more quickly than others, for example (Diener, 2000; Diener et al., 2006). Further our perception of wellbeing appears to fluctuate in relation to our age—levels of SWB shifting over the life course (Diener et al., 2006). In addition, 'wellbeing' is an assemblage of different aspects of life, and its components do not necessarily all progress in the same direction over time, again suggesting there is not simply one set point. People can feel more positive about one area of their life while feeling less positive about another, simultaneously. Satisfaction at work, for instance, does not necessarily result in similar feelings in our marital relationships (ibid.). As described above, SWB is also influenced by our society's ideals and norms. Aspirations for and experiences of wellbeing can be somewhat specific to the country you live in, for example, reflecting national differences in social and cultural values (Diener & Suh, 2000). This suggests that there is no universalised set point for SWB.

The impact of social and cultural values upon our conception and experience of SWB is also evidenced in studies that find differences in SWB between individualistic and collective cultures. Individualistic societies tend to value individualism and attention to the self. Collective cultures favour collective behaviours. Within wealthier individualistic cultures, feeling good about oneself has been highly correlated with positive life satisfaction, for example (Diener & Diener, 1996). In collective societies, self-esteem appears to play a less fundamental role in life satisfaction (ibid.).

This evidence concerning variation in our experience of wellbeing together suggests that our levels of SWB can alter and that this alteration may be long-lasting. Because who we are, when (in our lives), where and who with, influences both our wellbeing and our response to adjustments within it (Crivello, Camfield, & Woodhead, 2009, p. 53). This arguably validates both the shaping of interventions to increase SWB and the use of SWB as a measure of positive or negative change within our lives.

A Framework for Conceiving of Good from Woods

In GfW, SWB was selected as GfW's approach to measuring woodland wellbeing. While practitioners were supported to experiment with approaches that might conceive of and measure wellbeing in different ways, we encouraged each case study practitioner-researcher to include collection of self-reported wellbeing within their evidence collection. With conceptions of SWB at the heart of the project, GfW aimed to produce research departing from a shared starting place and sharing a common point of comparison.

Contextualising Data

The research suggesting that our responses to changes in SWB are shaped by our current circumstances, background and tendency to adapt were also understood to be significant, confirming that collection of evidence about who spends time in woodlands (their current social, economic and cultural context) and how long the effects of their activity last were vital in understanding woodland wellbeing.

Pilot GFW Domains and Indicators of Woodland Wellbeing

The GfW interpretation and framework of woodland wellbeing used during the project were formulated in three distinct ways:

- Categories (or 'domains') of health, happiness and associated indicators that people are experiencing good in these areas were drawn from existing SWB definitions and evaluative tools.
- These categories and indicators were also developed to correspond with the types of wellbeing practitioners suggested were the outcome of the woodland-based health and happiness services they delivered. This took place towards the start of case study research, building on emerging findings. The domains and indicators proposed described emotional, social, psychological, physical and biophilic areas of wellbeing.
- Throughout the project, evidence of woodland wellbeing collected in GfW case studies were compared with the following wellbeing domains: emotional, social, psychological, physical and biophilic, which were derived from existing wellbeing measures and earlier GfW findings. Where a good fit was made, practitioners aligned it with these domains and associated indicators. Where a fit couldn't easily be made, practitioner-researchers made suggestions for expansion of categories and indicators (explored in the case study evidence in following chapters and brought together in the conclusion, Chapter 11) (Fig. 3.1).

Emotional Wellbeing

The negative and positive affect or emotion-based aspects of SWB (theories of their impact and models of their measurement) contributed to GfW's conception of woodland wellbeing through a category or domain called 'emotional wellbeing'. GfW's interpretation of emotional wellbeing corresponds with experiences of positive or negative moods, feelings and emotions when spending time in the woods.

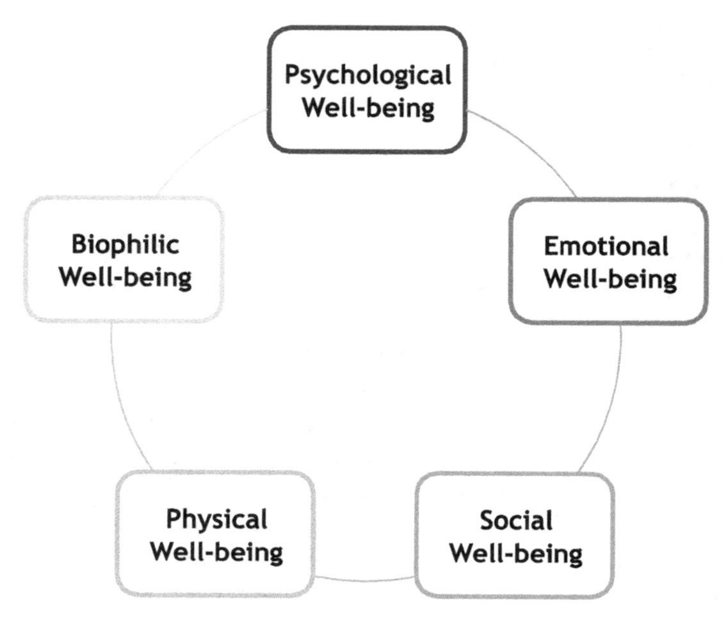

Fig. 3.1 GfW pilot wellbeing domains (*Source* Goodenough, 2015)

These affective outcomes need not be conceived of as intangible, fleeting impacts. Fredrickson's 'broaden and build' model of positive emotion suggests that experiences of good affect can grow our access to wellbeing (Fredrickson & Joiner, 2002). It predicts that, while negative emotions may limit people's behaviours, positive emotions can 'broaden' our thinking patterns, resulting in thoughts that are distinctively elastic, fertile and contemplative (ibid.). These types of thoughts can encourage us to experiment with new types of behaviour, activity and interpersonal connection (Fredrickson, 2013). In turn, these exploratory, novel projects contribute to our personal growth and development. This 'building' of personal assets (experience, flexibility, skills, social networks, etc.) also increases our capacity to access wellbeing in different areas, including more, stimulating, positive emotion (ibid., p. 16). These are the 'upward spirals' that positive emotions can effect (Fredrickson & Joiner, 2002).

Physiologically, experiences of positive emotion have been found to help recovery from the cardiovascular effects of stress (Fredrickson, 2013,

p. 12). The 'broaden' thought patterns of positive emotion have been associated with the ability to find meaning within adversity, helping us to manage challenges too (Fredrickson, 2013). They have also been associated with social connection, helping us to think in trusting and inclusive ways (ibid., p. 23). In sum, there is a weight of research suggesting that while frequently perceived as transitory, positive emotions can have far-reaching effects upon our health and happiness and should therefore form a part of evaluating woodland wellbeing.

Social Wellbeing

In individualistic societies, social relations play a more significant role in determining SWB than economic resources, while the opposite is true in collective cultures (Diener & Diener, 1996; Helliwell & Putnam, 2004; Jordan, 2008). The relatively high worth of social connection in wealthier countries has been argued to be the result of neoliberalist conditions. Through fostering individualistic behaviours (competition, home ownership, migration, flexibility and personal choice), rather than reinforcing the collective aspects of life (commitment to relatives, communities, civic responsibilities and membership of unions and associations), free market economies make social relations rarer and more valuable (Jordan, 2008). In terms of SWB, researchers suggest that social bonds and interpersonal relation in individualistic societies are particularly valuable resources of positive health and happiness (Helliwell & Putnam, 2004). A spouse, loyal friends, affable neighbours and sympathetic colleagues have all been identified as elements of social networks that beneficially influence wellbeing, including self-esteem, moods, diet, sleeping patterns and social integration (ibid.). Research shows that, as discussed, in relatively wealthy economies, levels of SWB can perhaps be best forecast by establishing the extent and vigour of a person's social associations (ibid.). It has been suggested, in fact, that the Easterlin complex may be a reflection of the influence of free market economics upon the shape of our social relations: an imbalance between our economic and social wellbeing. Collective networks and associated patterns of behaviour, such as trust and mutuality, can be described as aspects of 'social capital' (Helliwell & Putnam, 2004, p. 1436). This phraseology seeks to recognise that social networks and connections are of direct worth to their participants and their neighbourhoods (ibid.).

Many of the services delivered in woodlands and activities undertaken there have a social component and could conceivably have an impact on people's social capital—in terms of both increased social connection and cohesion and participation in behaviours associated with them (such as trust). In GfW, the value of associated social capital to wellbeing was acknowledged in a domain described as 'social wellbeing' and set of associated indicators of this type of wellbeing.

Psychological Wellbeing

With the use of SWB as a measure of societal progress was gaining currency in the UK, there was also increasing consent that such evaluations must take into account not only its 'hedonic' aspects, but also its 'eudaimonic' components. Eudaimonic wellbeing, though related to SWB and similarly subjectively assessed, is defined by distinctive sets of attitudes and behaviours (Keyes, Shmotkin, & Ryff, 2002; Ryff & Singer, 2006). While hedonic assessments focus on people's satisfaction, enjoyment and pleasure in life, as described above, eudaimonic accounts attend to how we are functioning and how much we are able to realise our potential (Michaelson et al., 2009). This distinction might be simplified as the difference between measuring how we are *feeling* and what we are *doing* (ibid.). This form of wellbeing is realised through directing our attention, interactions and activity towards satisfying our psychological needs. And possibly its effects may be more stable than the reactive emotional moods associated with hedonic wellbeing (Bauer, McAdams, & Pals, 2008). It is as context-specific as SWB, similarly varying in relation to factors including an individuals' age, character traits and sociocultural circumstances (Ryff, 1995, 2014).

There are several approaches to defining and measuring eudaimonic wellbeing. While different, the various models can be seen to converge in their description of a number of human psychological requirements whose fulfilment can help us to thrive (Vanhoutte, 2014).

Autonomy describes the need to feel we are self-determining and able to take independent actions consistent with our feelings and thoughts (Ryff, 2014; Ryan & Deci, 2004).

Competence or *Personal Growth* can be understood as a need to feel self-confident and effective in our actions and to find opportunities to demonstrate our proficiency to others (Ryan & Deci, 2004). It also encompasses the requirement to feel we are developing, realising our potential, becoming more self-aware and effective (Ryff, 2014).

Relatedness concerns a need to relate to, feel integrated with, express and experience understanding from others. Relatedness can be exhibited in desire to belong and feelings of security derived from feeling unified with others (Ryan & Deci, 2004).

Life Purpose describes the need for us to perceive our lives as having significance, purpose and positive progression (Ryff, 2014).

Environmental Mastery or Control concerns our need to feel we can control and manipulate our environment to meet our needs (Ryff, 2014).

Self-Acceptance is about our requirement to feel comfortable with ourselves and accepting of our personal qualities, positive and negative, and limitations (Ryff, 2014).

These needs to function positively could also be summarised as our requirement to be making sense: making sense of the world, making sense to ourselves and making sense to others (Ryan & Deci, 2004). Ryan and Deci (2004) suggest these needs are met through human behaviours that are aspirational, connected, engaged, educated, motivated, optimistic and purposeful. Positive functioning flows through our absorbing of events, perceptions and ideas, taking on of values and views and expression of these in behaviour, activity and relationships (Bauer et al., 2008, pp. 81–83). Research suggests that if negative events can be interpreted as contributing to this flow of development and growth (regarded as transformative challenges or a turning point), rather than disrupting it, positive functioning can flourish and resilience be achieved (Bauer et al., 2008; Keyes et al., 2002). This kind of optimistic re-telling of our life story can be seen as a strategy for achieving some of the needs described above. However, if difficult social and environmental factors cannot be translated as cathartic events in our individual narrative, then positive functioning is threatened (Bauer et al., 2008). These personal biographies of positive functioning are culturally specific, reflecting our sense of progress in relation to our society's values and expectations (ibid.).

Another aspect of eudaimonic experience relevant in the context of woodland wellbeing is our need for 'vitality'. Vitality describes our experience of personal physical and mental energy—an energy bridging our PWB and physical wellbeing (Ryan et al., 2010). For example, positive vitality is associated with other aspects of PWB that can support our growth and development (ibid.). Equally, subjective feelings of vitality can be correlated with our awareness of our bodily functions and symptoms. Good health should be expected to contribute to us feeling vital (ibid.). Vitality is our fuel for our thriving, and our thriving can contribute to feelings of vitality (ibid.). Sometimes felt as a sense of aliveness, being refreshed and enthused, a lack of vitality conversely can leave us feeling depleted and disengaged (Ryan & Frederick, 1997).

These various components of eudaimonic happiness are reflected within the GfW category of 'psychological wellbeing'. This was anticipated, together with emotional wellbeing, as the area in which findings regarding positive and negative mental health outcomes would show up in our data. As Abdallah and colleagues (2008) point out, it is possible to experience positive emotions while having poor PWB (by temporarily altering our state of mind by stimulant use, for instance) and to feel negative while achieving personal development and growth (through grief, e.g. as explored above).

Physical Wellbeing

Our physical wellbeing can also be assessed using subjective accounts—self-reported assessments of health behaviours and perceived physiological health (Abdallah et al., 2008).

Physical health and activity appear to be closely linked to other aspects of our welfare (Abdallah et al., 2008; Fredrickson, 2013; Ryff, 2014). In particular, positive health habits (exercise, healthy eating, sleeping well, for example) can be associated with good social, emotional and PWB (ibid.). Some analyses suggest that levels of emotional, PWB and social wellbeing can be drivers of good health behaviours—and vice versa—that physical health can contribute to emotional, social and PWB resources through 'feedback loops' (Abdallah et al., 2008; Fredrickson, 2004).

Physical activity, because of practices and behaviours associated with it, appears not only to have physiological impacts, but is frequently accompanied by psychological (control and competence, for example) and social (friendships and teamwork, for instance) outcomes and pathways (ibid.).

Our physical bearing is an area where emotional wellbeing appears to combine with our physical state as embodied emotion. Research suggests that our bodies can reflect our experience of positive or negative emotion. Positive emotion, for example, may be manifested in a more open torso and extended neck—bodily expansiveness—than negative or neutral emotional states (Fredrickson, 2013).

Biophilic Wellbeing

As described in Chapter 2, there is growing theory and evidence suggesting not only that nature positively impacts human health and happiness, but that humans have a need for nature. Our biophilic needs and experiences have been explored through the complementary theories that understand them to be adaptive reflexes belonging to our evolution in nature (Chapter 2). Together these suggest that natural environments could (amongst other effects): cause us involuntary arousal of interest, positive or negative emotional responses (Ulrich et al., 1991; Wilson, 2009), support recovery from stress (ibid.) and restore our capacity to pay attention (Kaplan & Kaplan, 1989). As discussed in Chapter 2, experimental studies have confirmed that green views, nearby nature and forest settings can indeed have psychologically and physiologically supportive effects and a range of associated positive outcomes for health and wellbeing.

Biophilia may also be demonstrated in our felt sense of having a 'connection' to nature (Lumber, Richardson, & Sheffield, 2017). Nature connection concerns perceptions and experiences of feeling part of the natural world and valuing it beyond its material provisioning of human needs (ibid.). This feeling of relatedness is subjectively experienced and can be measured through our individual sense of it (Richardson et al., 2019). Like the other forms of SWB explored above, our experience of it appears to vary in relation to sociocultural factors including our background, life experiences, knowledge, character traits and life-stage (Lumber et al., 2017; Richardson et al., 2019). For example, evidence points to a drop in nature connection during the teenage years or 'adolescent disconnect' (Richardson et al. 2019, p. 13). Richardson et al. (2019) conjecture that this is related to the many developmental and situational changes occurring during this period, but we could also speculate whether we are yet employing appropriate methods or questions to investigate when, where and how young people relate to the natural world (explored in Chapter 9).

The GfW concept of biophilia and nature connection as part of woodland wellbeing was developed in a category called 'biophilic wellbeing' (Table 3.1).

Table 3.1 GfW pilot indicators of wellbeing

Psychological wellbeing—positive functioning	Feelings of: being in control, competent (and seen by others to be competent), energetic, purposeful, developing oneself, connecting with others through shared beliefs and outlook, secure with personal limitations
Emotional wellbeing	Experiencing: positive emotions and moods, absence of negative emotions and moods, feeling even-tempered, relaxed, optimistic about the present, optimistic about the future
Social wellbeing	Feelings of: being confident, accepted, safe and supported *within and through* social relationships, supporting others through social relationships
Physical wellbeing	Feelings of: physical health, confidence in and enjoyment of physical activity
Biophilic wellbeing—connection to nature	Feelings of: closeness to the natural world, being engaged in a relationship with nature

Source Goodenough (2015)

This chapter has drawn attention to some of the complexity of the wellbeing concept, its broad definition and variations in the ways that individuals may experience different aspects of it—reflecting our diversity (biographical, cultural, psychological, biological, etc.). The GfW framework for exploring woodland wellbeing was deliberately developed with broad domains and indicators and an invitation for practitioner-researchers to test and expand it in order to allow inclusion of the range/subtlety of experiences that might be relevant. How woodland wellbeing sometimes eluded and expanded these ideas is explored further in the empirical chapters following Chapter 4 and within our concluding chapter. Our next chapter explores what approaches supported practitioners in developing research designs that helped collect evidence within and beyond the definitions that had been standardly included in the literature.

References

Abdallah, S., Steuer, N., Marks, M., & Page, N. (2008). *Well-being evaluation tools: A research and development project for the Big Lottery Fund, New Economics Foundation.* Retrieved from http://www.biglotteryfund.org.uk/wellbeing_evaluation_tools.pdf.

Adler, A., & Seligman, M. E. P. (2016). Using wellbeing for public policy: Theory, measurement, and recommendations. *International Journal of Wellbeing, 6*(1), 1–35.

Allin, P., & Hand, D. J. (2017). New statistics for old?—Measuring the wellbeing of the UK. *Journal of the Royal Statistical Society Series A, 180*(1), 3–43.

Bacon, N., Brophy, M., Mguni, N., Mulgan, G., & Shandro, A. (2010). *The state of happiness: Can public policy shape people's wellbeing and resilience?*. London: The Young Foundation.

Bauer, J. J., McAdams, D. P., & Pals, J. L. (2008). Narrative identity and eudaimonic well-being. *Journal of Happiness Studies, 9*(81), 81–104.

Brekke, K., & Howarth, R. B. (2002). *Status, growth and the environment: Goods as symbols in applied welfare economics.* Cheltenham: Edward Elgar Publishing.

Crivello, G., Camfield, L., & Woodhead, M. (2009). How can children tell us about their well-being? Exploring the potential of participatory research approaches within Young Lives. *Social Indicators Research, 90,* 51–72.

Defra. Retrieved from http://www.defra.gov.uk/sustainable/government/what/priority/wellbeing/common-understanding.htm.
Diener, E. (2000). The science of happiness and a proposal for a national index. *American Psychologist, 55*(1), 34–43.
Diener, E., & Diener, C. (1996). Most people are happy. *Psychological Science, 7*(3), 181–185.
Diener, E., & Suh, E. (2000). National differences in subjective wellbeing. In D. Kahneman, E. Diener, & N. Schwarz (Eds.), *Well-being: Foundations of hedonic psychology* (pp. 434–452). New York: Russell Sage Foundation.
Diener, E., Lucas, R. E., & Napa Scollon, C. (2006). Beyond the hedonic treadmill: Revising the adaptation theory of well-being. *American Psychologist, 61*(4), 305–314.
Dodge, R., Daly, A., Huyton, J., & Sanders, L. (2012). The challenge of defining wellbeing. *International Journal of Wellbeing, 2*(3), 222–235.
Dolan, P., & White, M. P. (2007). How can measures of subjective well-being be used to inform public policy? *Perspectives on Psychological Science, 2*(10), 71–85.
Dolan, P., Layard, R., & Metcalfe, R. (2011). *Measuring subjective well-being for public policy.* Retrieved from http://eprints.lse.ac.uk/35420/1/measuring-subjective-wellbeing-for-public-policy.pdf.
Frank, R. H. (1999). *Luxury fever: Why money fails to satisfy in an era of excess.* New York: The Free Press.
Fredrickson, B. L. (2004). The broaden-and-build theory of positive emotions. *Philosophical Transactions of the Royal Society, 359,* 1367–1377.
Fredrickson, B. L. (2013). Positive emotions broaden and build. In P. Devine & A. Plant (Eds.), *Advances in experimental social psychology* (Vol. 47, pp. 1–53). Burlington: Academic Press.
Fredrickson, B. L., & Joiner, T. (2002). Positive emotions trigger upward spirals toward emotional well-being. *Psychological Science, 13*(2), 172–175.
Goodenough, A. (2015). *Social cohesion and wellbeing deriving from woodland activities: Good from woods* (A Research Report to the BIG Lottery).
Helliwell, J. F., & Putnam, R. D. (2004). The social context of well-being. *Philosophical Transactions of the Royal Society London B, 359,* 1435–1446.
Horowitz, D. (2018). *Happier? The history of a cultural movement that aspired to transform America.* New York: Oxford University Press.
Jordan, B. (2008). *Welfare and well-being: Social value in public policy.* Bristol: The Policy Press.
Kahn, P. H. (1999). *The human relationship with nature: Development and culture.* Cambridge: The MIT Press.

Kahneman, D., & Krueger, A. B. (2006). Developments in the measurement of subjective well-being. *Journal of Economic Perspectives, 20*(1), 3–24.

Kaplan, S., & Kaplan, R. (1989). *The experience of nature: A psychological perspective.* Cambridge: Cambridge University Press.

Keyes, C., Shmotkin, D., & Ryff, C. (2002). Optimizing well-Being: The empirical encounter of two traditions. *Journal of Personality and Social Psychology, 82*(6), 1007–1022.

Louv, R. (2005). *Last child in the woods.* New York: Workman Publishing.

Lumber, R., Richardson, M., & Sheffield, D. (2017). Beyond knowing nature: Contact, emotion, compassion, meaning, and beauty are pathways to nature connection. *PLoS One, 12*(5), e0177186.

Michaelson, J., Abdallah, S., Steuer, N., Thompson, S., & Marks, N. (2009). *National accounts of well-being: Bringing real wealth onto the balance sheet.* Retrieved from http://cdn.media70.com/national-accounts-of-well-being-report.pdf.

Mulholland, H., & Watt, N. (2010). David Cameron defends plans for wellbeing index. *The Guardian.* Retrieved https://www.theguardian.com/politics/2010/nov/25/david-cameron-defends-wellbeing-index.

O'Brien, L. (2009). *Wellbeing, forestry and ecosystem services: A discussion paper.* Forest Research. Retrieved from http://www.forestry.gov.uk/pdf/Forestry_and_wellbeing_discussion_paper.pdf/$FILE/Forestry_and_wellbeing_discussion_paper.pdf.

Richardson, M., Hunt, A., Hinds, J., Bragg, R., Fido, D., Petronzi, D. ..., & White, M. (2019). A measure of nature connectedness for children and adults: Validation, performance, and insights. *Sustainability, 11*(12), 3250.

Ryan, R. M., & Deci, E. L. (2004). An overview of self-determination theory: An organismic dialectical perspective. In R. M. Deci & R. M. Ryan (Eds.), *Handbook of self-determination research* (pp. 3–36). Suffolk: The University of Rochester Press.

Ryan, R. M., & Frederick, C. (1997). On energy, personality, and health: Subjective vitality as a dynamic reflection of well-being. *Journal of Personality, 65*(3), 529–565.

Ryan, R. M., Weinstein, N., Bernstein, J., Warren Brown, K., Mistretta, L., & Gagné, M. (2010). Vitalizing effects of being outdoors and in nature. *Journal of Environmental Psychology, 30*(2), 159–168.

Ryff, C. (1995). Psychological well-being in adult life. *Current Directions in Psychological Science, 4*(4), 99–104.

Ryff, C. (2014). Psychological well-being revisited: Advances in the science and practice of eudaimonia. *Psychotherapy and Psychosomatics, 83,* 10–28.

Ryff, C. D., & Singer, B. H. (2006). Best news yet on the six-factor model of well-being. *Social Science Research, 35*(4), 1103–1119.

Thompson, S., & Marks, N. (2008). *Measuring well-being in policy: Issues and applications.* New Economics Foundation.

Ulrich, R. S., Simons, R. F., Losito, B. D., Fiorito, E., Miles, M. A., & Zelson, M. (1991). Stress recovery during exposure to natural and urban environments. *Journal of Environmental Psychology, 11*(3), 201–230.

Vanhoutte, B. (2014). The multidimensional structure of subjective well-being in later life. *Journal of Population Ageing, 7*(1), 1874–7876.

White, S. (2018). *Moralities of wellbeing* (Bath Papers in International Development and Wellbeing; No. 58/2018, pp. 1–17). Bath: Centre for Development Studies, University of Bath.

Wilson, E. O. (2009). Biophilia and the conservation ethic. In D. J. Penn & I. Mysterud (Eds.), *Evolutionary perspectives on environmental problems* (pp. 249–258). New Brunswick and London: Aldine Transaction.

4

Assessing the Affective in Active Spaces

In the last chapter, we deconstructed the complexity of what 'woodland wellbeing' might be and explained the Good from Woods framework that enabled practitioner-researchers to identify nuance in types of wellbeing derived from woodland-based activity. We now turn to how practitioner-researchers selected appropriate research methods to capture these affective outcomes so that they and their organisations could make practical use of the evidence. The GfW project research team and practitioner-researchers collaborated in trialling and testing research design, management and delivery, identifying factors that increase success and the kinds of approach that best delivered findings in specific woodland activity contexts. Through this collaboration within the GfW project, a practical guide to preparing, planning and analysing research focused on woodland wellbeing was designed and produced. This chapter explores how wellbeing research is influenced by the woodland context, and the people, sites and practices involved in activities there. It critiques the congruence of certain research tools with active, outdoor contexts, illustrated by experience from the Good from Woods project. The learning from this process is also captured in a free online resource called the **Good from Woods research toolkit** (GfW, n.d.).

A. Goodenough and S. Waite, *Wellbeing from Woodland*,
https://doi.org/10.1007/978-3-030-32629-6_4

Why Measure Woodland Wellbeing?

The GfW project aimed to develop the research capacity of woodland-based service providers in response to the rapid growth of this sector of outdoor activity. As described in the introduction, reforms to public health service delivery have created new opportunities for those working in the woodland sector to deliver wellbeing services inspired by and benefiting from this natural context. At the same time, they have established the need for service providers to be able to identify and articulate factors that make their service successful and demonstrating the degree of this success to commissioners and funders of services and service users. However, subjective wellbeing is, as we have already discussed, a highly personal and affective state and one that it is often difficult to evidence. There are standardised measurement tools available such as the Warwick-Edinburgh Mental Health and Wellbeing Scale, but these sometimes don't sit well with the mobile nature of woodland activity (Waite & Waters, 2019) and the material character of many of the outdoor sites mean that the time and space to do online- or paper-based surveys before, during or after may be constrained or entirely absent. Furthermore, many of the communities that outdoor health and wellbeing interventions are targeted towards may struggle to express their sense of wellbeing and its origin using quantitative self-report measurement approaches.

Who Should Measure Woodland Wellbeing?

While many interventions are evaluated separately as a summative post-activity action (Waite, Rutter, Fowle, & Edwards-Jones, 2015), Good from Woods was premised upon the involvement of practitioners in researching, identifying and implementing best practice. The intention was that practitioners delivering natural health and happiness services in this environment would be ideally placed to develop research approaches to evidencing woodland wellbeing that were firmly grounded within this context and responsive to the circumstances of particular client groups. Developing their research skills would also enable a sustainable evaluation approach within these organisations, supporting the evidencing of benefits in a quality assured manner without reliance on external research expertise. Service providers were supported within Good from Woods through training and mentoring to become action researchers, expert in their own field of delivery and best placed to sustain achievements and implement change. The overarching concept of wellbeing as a potential outcome from their work held the diverse cases in common and facilitated cross-case learning and peer support. Clearly, there are circumstances in which an external evaluation or research study may be preferable, but an action research approach meant that Good from Woods' aim of upskilling practitioners in research methods could be actively promoted alongside the collection of data grounded in local social and environmental knowledge.

Action Research

Action research works towards the identification of opportunities for positive change and further research through people researching their own lived experiences (McNiff, 2012, p. 3). The theoretical, ideological and practical roots of this approach are varied and cross-disciplinary (Charles & Ward, 2007, pp. 2–3). The action research model brings together two behaviours sometimes perceived as separate impulses—that of the scholar and the activist—promoting the combination of intellectual activity and practical outcomes (Somekh & Zeichner, 2009). Action research approaches are frequently aimed at democratising who undertakes and participates in research, giving voice to perspectives, knowledge and ideas that might otherwise remain unheard and unacknowledged (Carr & Kemmis, 1986, p. 164; Gaventa & Cornwall, 2008, p. 176). Action researchers are active members of the community being investigated, rather than detached, visiting observers. Their ties to the field of research can provide them privileged access to expertise and insight (Charles & Ward, 2007, p. 5). The action research literature affirms a collaborative approach to the gathering and understanding of data, a shared process that can raise awareness of local circumstances and increase the potential for altering these (ibid., 2007, p. 2). These key ideas within action research practice (the practical value of research, participative problem solving and action) offered a good fit with enabling service providers to capture and describe the value of their service.

Starting Points

At an early point in the project, we used a questionnaire and workshop to establish practitioner-researchers' views of what it is to be a researcher (discussed further in the next section). These included visions of people in white coats, static at computers and importantly, always indoors. Research was seen as something removed from the activity that it studied. There was a strong perception amongst several practitioners that 'research' was not something that they did and that they had deliberately chosen a job that in contrast allowed them to be physically active in outdoor settings. The research training therefore began by drawing attention to underlying research skills that practitioners used regularly in their daily life, such as observation, dialogue and reflection and supporting the practitioner-researchers in becoming more aware of their position in the field of research and possible preconceptions. The project research team had reviewed a book on Real World Research (Goodenough & Waite, 2012) around that time that was useful in highlighting how the gap between research and practice might be overcome. We thought that the mantra of 'systematically', 'sceptically', 'ethically' (Robson, 2011, p. 15) could help new researchers develop commonsense inquiry into a more rigorous research approach while recognising and valuing their existing knowledge and skills. Part of that process would be to make practitioners more aware of their expectations from their practice and a further workshop was held that elicited understanding of what both woodland and their activities might bring to wellbeing, including potential outcomes and the factors to which they might be attributed.

As described, GfW was a collaborative research project, partnering practitioner-researchers with a research team: the project-researcher, based within a third-sector organisation, and a mentor at Plymouth University. Forms of support practitioner-researchers accessed from the research team included frameworks for guiding research design, introduction to research approaches and tools, and individual advice and guidance. Partners sometimes experienced frustration with differences between working practices (Goodenough, 2015). However, each collaboration achieved valued learning, skills and new relationships through co-production (ibid.). For some practitioner-researchers, partnership with the research team, despite challenges, was important to plan and achieve their research goals.

> Having a person outside the organisation to support the process really helped the research to bear fruit... without someone guiding the way I think I would have given up somewhere in the middle [R7].

Researchers also collaborated in sharing learning across the group of case studies commissioned under GfW through meetings and online discussion. Sharing learning with others evaluating service delivery could be an important source of mutual support and guidance (Fig. 4.1).

GfW's action research cycle. Its cyclical design is intended to move beyond explanations of local circumstances, towards advancing changes within them (Carr & Kemmis, 1986, p. 194).

Fig. 4.1 GfW's action research cycle (*Source* Goodenough, 2015)

Tidying Up or Getting Messy?

Practitioners deciding to take on wellbeing research were concerned about achieving a balance between research, evaluation and delivering woodland activity. A research focus was sometimes perceived as a possible conflict with their existing work—creating and managing the conditions in which participants can benefit from being in the woods. Similarly, some felt that capturing data from participants might require traits and skills not normally used in their woodland role. Research was sometimes perceived as intellectual, indoors, scientific, quantitative and bureaucratic. The strength of these ideas sometimes influenced who practitioners understood could do research, where and how, but also which sort of research results might have value.

Practitioners identified several abilities they felt transferred between and bridged data collection and service delivery, including observation, communication, reflection and organisation skills. Others, like flexibility and leadership, they connected more closely with service delivery.

> Sensitivity to group and individual dynamics, good observation skills, concise and accurate recording skills, listening and open, non-judgemental attitude (P2—researcher skill).

> Organisation, forward planning, computer skills. Good communication. Ability to gather and make sense of data…Friendly, approachable and persuasive to encourage people to engage with the research (P3—researcher skills).

> The ability to put people at ease in an unknown setting is useful, as well as a clear awareness of their needs at all times. Flexibility is essential as families can be unpredictable, as can the environments I use for learning. Creativity allows for changes of plan and making the best of any situation (P1—woodland practitioner skills).

> Practical outlook and willingness to get mucky and have a go. Leadership and confidence with groups. Listener, problem solver, solution focused (P6—woodland practitioner skills).

Strikingly, the research role was associated by service providers with a sort of tidying—bringing focus, order and accuracy to the fore. In contrast, woodland-based work was seen as somewhere you might get messy, both literally and figuratively, interacting strongly with the environment and people. Service providers tended to see practitioners as firmly embedded in their context while imagining a researcher's clarity to be tied to a more emotionally neutral approach. In this sense, training for research could be conceived as training emotions. The tidying up of the practitioners' emotions paralleled by a reification of affective wellbeing as something that could be neatly parcelled into categories.

This understanding of research as needing to be emotionally neutral, orderly and neat caused concerns for practitioner-researchers during the process of their research. Feeling messily embedded in the community and activities that were being researched sometimes felt inimical to those goals. For example, their wide-ranging knowledge of their woodland-based provision made determining a clearly defined area of focus sometimes daunting. Some practitioner-researchers were attracted by the idea that a standardised, impartial research tool could be developed or used. However, these also brought complexities in design or transfer to the scale and (ir)regularities of woodland-based service delivery and its participants. For many practitioner-researchers, woodland-based research design, data collection and analysis got 'messy' at some point, before clarity could be achieved (Cook, 2009).

> there were those times… it reminded me a bit of what's that bit you get to when you're giving birth? What do they call that bit, transition?…There were times like that! When you thought "oh I just really don't know what I'm doing here, or why I'm doing it or where I'm going?"

> It is very interesting what a roller coaster this research project has been - I do hope that you are able to capture the process that we are going through [experiential learning] as much as data and indicators! [R7]

> I will now go through and tidy it [the data reporting]up… edit it into real English…I feel that my brain has expanded which is a great thing…Right,

> now I'm off to find some livestock to organise and care for. Something tangible to do [R8].

Experiential learning gained during the research and evaluation process often proved central to practitioners' conceiving what a comfortable research role, integrated within their existing delivery, might look and feel like. Sometimes feeling messy could in fact signal a 'transition' or turning point in approach or understanding. The evidence suggests that it is useful to find an approach that can be invested in and practised over a period of time, while acknowledging that the process may be untidy at times and require adjustments or more consideration. While many practitioners came to understand the value of research getting complicated, the value of experiential learning while designing, collecting and analysing data is often under emphasised and unexplored in research reporting (ibid.). The support of the project-researcher could be crucial, acting as a kind of midwife in assisting the birth of new insights and encouraging sustained involvement when the tensions between delivery and research were taut.

Though a tension for some, practitioner-researchers' position within the field (or wood!) provided distinct advantages when they felt confident to draw on their existing understandings and relationships within this research setting.

> Being personally involved with [initiative] really helped me to 'get into' the research, and I feel that the process would have been more laborious had there been 'outside' researchers - it was very valuable that we all started with a shared understanding of the [initiative] [R21].

> my good existing relationship with the facilitators [of woodland activities] and participants…my familiarity with [initiative] and the woodland [were factors facilitating research] [R2].

Practitioner-researchers also adopted research practices most confidently when they could see their existing skills and abilities were relevant to it.

> I took on the role of facilitating focus groups and interviewing [initiative] members. I have years of experience facilitating people providing information about themselves in groups and individually, so I felt very at ease in this role [R20].

> I have a background in project management, so took on the role of co-ordinating the team effort - making sure we kept to the [research] timetable [R21].

Where practitioner-researchers chose a more 'external' research approach, it sometimes made them feel at odds with their normal methods of delivery, service users, colleagues and even their organisational ethos.

> I feel that I could have gotten better answers [from service-users] if I had been part of the [woodland activity] facilitation team and then asked questions after being part of the team that enabled the experience [R8].

> It's challenging being at [woodland service delivery organisation] with a culture where we do lots and don't sit and reflect and research all that much…I feel self-conscious [undertaking research] as maybe we don't respect this kind of 'work'…it would be better if the research was less explicit but more deeply interwoven in the [woodland] activity, this feels like best practice [R2].

However, this discomfort could also lead to insights that might not have been so apparent in the thick of delivery:

> …took a lot of still photos and made notes - sitting as an observer whilst [colleagues] led. This felt very awkward; next time I'd like to be able to lead some reflective activities…morning session felt quite chaotic; interesting to observe rather than try and put right [R7].

Smaller and/or less established organisations were less likely to plan how and where woodland wellbeing research results would be used within their organisation. Sometimes it was unclear who would supervise or be accountable for the success of data collection and use. Flatter management structures sometimes resulted in practitioner-researchers feeling less able to request support from peers and managers. Accustomed to defining themselves as active 'doers', needing limited supervision, some practitioner-researchers were reluctant or struggled to alter this image amongst colleagues. Perceptions of differences between 'thinkers' and 'doers' (intellectual and practical capacities) along the lines of the researcher and the practitioner influenced the ease of research delivery and use of its results.

Less established and smaller scale initiatives were also more likely to experience shifts in priorities, commitments, funding and staffing, which threatened supervisory, peer and practitioner-researcher focus upon data collection and management. However, sudden shifts in aims, obligations, financial support and personnel could also impose within larger organisations and challenge embedding of research cultures. Without strong management and peer understanding of and support for research goals, such changes can diminish the capacity of organisations and practitioners to carry out woodland wellbeing research and capitalise on its value internally.

What seemed to work most successfully, and was aspired to by many practitioner-researchers, was a research approach:

- That integrated research with existing familiar behaviour and skills;
- Where research aims were understood and respected by colleagues and managers;
- That practitioner-researchers felt achieved clear and valued outcomes on behalf of their organisation and themselves.

These conditions for success were sometimes only fully appreciated following negotiation of the challenges and opportunities of gathering and interpreting woodland wellbeing data. The experience and process of carrying out research was often key in practitioner-researchers' development of a blueprint for integrating research and evaluation into their organisations and their own ongoing work.

Our evidence suggests that it is important that woodland-based research and evaluation methods are a good fit with practitioners' existing working context and identities. Choosing approaches and methods that suit your audience for the results is important, but practitioners also need to feel confident and comfortable with data collection, analysis and reporting approaches and their objectives.

Key challenges and support for integration of research and practice
Challenging to integrated research/evaluation:

- Perceived/felt disjuncture between thinkers and doers – theoretical understanding and practical abilities
- Lack of clarity over research goals and objectives within organisation
- Low engagement with research methods and goals across organisation: fragile micro-research culture
- Unclear responsibility for achievement of research outcomes within organisation
- Shifts in priorities, commitments, funding and staffing
- Low integration of research approach with existing service delivery methods
- Feeling overwhelmed by research tasks and data: Getting 'messy'.

Supportive of integrated research/evaluation:

- Clearly shared understanding of research goals and objectives
- Clear planning for use of findings
- Involvement of multiple staff members within the research: sustaining micro-research culture
- Established lines of responsibility for achieving research outcomes: sustaining micro-research culture
- High integration of research approach with service delivery methods
- Recognising and valuing experiential learning and 'getting messy'.

Adapted from Goodenough (2015, p. 24)

Groundwork for Building a Holistic Picture of Health and Happiness Outcomes

The next section introduces some of the practical, reflective and theoretical strategies that supported practitioner-researchers to establish methodologies that felt grounded in the woodland context: people and place.

Practitioner-researchers initiated their research design and data collection process by identifying stakeholders, other than service users, in their service provision. Stakeholders could include: commissioners of the service; funders; institutions, organisations, health and social care professionals referring or supporting users; and users' families and carers. Stakeholders were a key source of evidence, able to suggest the outcomes they anticipated service users achieving from their service use. Collecting these views was a practical task, introducing the project to stakeholders and supporting them to describe the distinctive dimensions of woodland wellbeing they associated with service delivery. It was also reflective work, helping practitioner-researchers become aware of and consider contrasts between their conception of woodland wellbeing and that of others. Capturing stakeholders' perspectives could also help practitioner-researcher's establish what audiences for their findings were interested in knowing and the kinds of evidence they might find convincing. Finally, it provided a way of establishing the *purposes and expectations* of service provision. This was a variable that practitioner-researchers might identify as influential in how a service was aiming at or achieving woodland wellbeing outcomes. It was the first factor, in a series of variables that practitioner-researchers were encouraged to look for as potentially related to outcomes within their setting. As well as exploring expectations and purposes of sessions (why), practitioners were invited to explore the influence of: people involved (who); the wooded setting (where); the activities (what); and the ethos and approaches used within delivery (how) (Waite, Bølling, & Bentsen, 2016).

Purposes, people, places, activities and practices: variables influencing health and wellbeing outcomes

Practitioner-researchers were encouraged to establish a comfortable research role and methods with which to explore variables that might influence wellbeing including:

- expectations and purposes of sessions (why)
- the people involved (who)
- the wooded setting (where)
- the activities (what)
- ethos and approaches used (how).

Adapted from Waite et al. (2016).

Stakeholder mapping made audible a wide range of voices (age, position or level of approval aside) and highlighted convergence and difference amongst viewpoints, allowing persistent themes to emerge and influence the ongoing focus of the research. Practitioner-researchers could then use this groundwork to help guide how they would gather findings from service users, some focusing a high proportion of data collection on one aspect of wellbeing for example, others following a much more open-ended enquiry.

Practitioner-researchers were encouraged to keep a reflective diary, journaling the research process. This written record was not intended to communicate the research process to others (although they were shared with the research team), but principally to log their own observations and reactions, noting what worked or did not, the unexpected or fortuitous, hunches followed, emotions and learning experienced (Newbury, 2001). Research diaries and reflection supported practitioner-researchers in reviewing how effectively their research design fitted their context and aims and how adjustments to this strategy might be achieved.

For example, several practitioner-researchers with arts backgrounds or interests developed expressive arts derived techniques.

> ...just held the arts-based research workshop in the wood [aimed at establishing how co-owners of community woodland benefit from ownership]...format seemed to work well – based on a role play scenario of

> developers wanting to build on the land, and reporters from the local newspapers interviewing members and gathering illustrations for articles…the children expressed their views very confidently and with real passion….my hunch is that talking to people whilst in the wood will have had an impact on the kinds of things they talk about and how they express themselves…I think it would have been much less successful talking to children in a neutral venue [P-R].

These more experimental strategies could feel particularly comfortable and engaging for practitioner-researchers who designed them and were interested and invested in what evidence might be generated. A disadvantage, however, was that some of the products of creative approaches (artwork, photographs or objects, for example) were sometimes less easy to analyse or report than text or numbers. On the other hand, these methods captured aspects of woodland wellbeing that service users might not translate into reflections or statements about health.

> …I felt that the coding couldn't capture the 'atmosphere' of people's comments – the vitality, the enthusiasm, the body language, the activity that was going on at the time (drawing, playing) – it's only about the words. Hopefully the drawings people did and the photos…will enable us to fill out what we get from the transcripts. It's a shame that we haven't got more time to do other workshop-type research activities which can capture more subtle stuff [P-R].

One practitioner, confident in developing creative data collection approaches, decided to take a further step of acknowledging woodland as a stakeholder in the research and a source of evidence. This recognition of the non-human world as a material force within woodland wellbeing goes further than attempting to capture its often intangible biophilic effects. It inspires attempts to find ways of recording and representing the entanglement of biological and biophysical processes within human experience and the impact of our encounters with other species (Kohn, 2013; Melson, 2013; Myers, 2015). In practice, it saw the practitioner-researcher observing how human behaviour was connected with nature's activity (e.g. human attraction to particular types of tree, flora, woodland material or topography and patterns of response to them), what woodland recorded of human–nature interactions (torn branches, stripped bark, bare earth, for instance) and the positive and negative aspects of these transactions for each species (see Chapter 9).

Standardisation in Measuring Woodland Wellbeing

One of the first places that the GfW project and practitioners looked for research tools and approaches was amongst existing standardised health and wellbeing measures. Designed by drawing on theoretical understanding, these measures are piloted and validated to ensure their interpretation by users is generally consistent, that the tool measures what it is meant to, and results correspond with those produced by similar instruments. In the case of measuring subjective wellbeing, these tools aim at reflecting something more 'stable' than fleeting mood and satisfaction, while remaining 'sensitive' enough to capture changes in these feelings over time (Pavot & Diener, 1993, pp. 165–167).

Within the UK health system, different services may employ tools suited to their service setting, users and style of interventions. However, there are also growing numbers of measures commissioned outside of a clinical context and for use beyond that environment. For example, there are a wide range of surveys (trialled to varying degrees) available for quantifying changes within mental wellbeing that reflect investment in our health and happiness from within the conventional health sector and beyond (such as British Broadcasting Corporation [BBC] and Office for National Statistics [ONS] tools) (Taggart & Stewart-Brown, n.d.).

During the lifetime of GfW, UK public services, NGOs, the third sector and academia have all contributed to the development of new health and happiness scales. This not only reflects growing interest in the concept of wellbeing, but also wider interpretations of what contributes towards it. For example, the New Economics Foundation has been highly influential in their inclusion of a wider range of health dimensions within subjective wellbeing measurement, including physical, social and psychological fitness. Producing surveys incorporating these domains of health and happiness, they have also promoted this broader conception through 'the five ways to wellbeing' framework—'Connect, Be active, Take notice, Keep learning and Give' (Abdallah, Main, Pople, & Rees, 2014, p. 5). Designed as a simple shortcut to recalling five key actions that support different components of wellbeing, it mirrors UK public health messages about remembering to eat your 'five-a-day' (fruit and veg) (ibid.).

As part of this expansion of subjective wellbeing measurement, instruments reflecting the role of natural environments in supporting health and happiness have also been developed. At the magnitude of entire countries, The New Economics Foundation developed 'The happy planet index', aimed at estimating a nation's collective wellbeing by factoring in crucial aspects of environmental and economic sustainability. In terms of personal wellbeing, both the Royal Society for the Protection of Birds (RSPB) (Bragg, Wood, Barton, & Pretty, 2013) and Natural England (Hunt et al., 2017) have commissioned the design of tools appropriate to capturing evidence of adults' and young people's connection with nature.

GfW practitioner-researchers explored the potential for using several standardised survey methods for capturing aspects of woodland wellbeing including one designed by the New Economics Foundation on behalf of the Big Lottery and the 'Emotional Literacy Checklist'. Both practitioners and the research team were attracted to the idea that these measures might transfer to a woodland context, perhaps reflecting ideas around the identity of researchers discussed above. As described, notions of researchers being detached, objective and able to impose order on the field, subjects and findings of research circulated in practitioner-researcher communities. Likewise, these ideas were also present in professional academic networks. For example, when introduced to the project, academic audiences frequently queried how research questions, methods and tools developed within the field could be balanced and robust enough to have impact beyond it. Some predicted that dependency on qualitative, non-standardised methods would limit its influence beyond its immediate context. It was suggested if its questions and methods were entrenched in local ideas and relationships, its findings could not be compared with or translated to other settings or approaches. Practitioner-led research was also seen as possibly tautological, establishing what it desired to see prior to enquiry and then only seeing that evidence.

In practice, however, practitioner-researchers found there were both barriers and opportunities in using existing, standardised measurement of wellbeing in an outdoor setting and with some of the communities that outdoor health and wellbeing interventions are currently targeted towards.

Standardised surveys presented several challenges for some woodland wellbeing providers. Not all measures are free, and additional funds might be needed to buy the most appropriate measure. Some tools require repeated delivery at set time-points and many completions for conclusions about patterns of change to be robust, when service users may only attend on an irregular or one-off basis depending on the style of activity. The mobile nature of many woodland activities could also make sedentary standardised survey methods counterproductive to capturing immediate embodied responses.

Further, some service users appeared reluctant to complete surveys, perhaps unsure as to whether the answers would affect their relationship to service providers, whether there was a wrong or right answer or whether they wanted to share the information being requested. This last issue may have arisen during other forms of data collection but was perhaps most obvious in this instance. These and other ethical issues such as what would happen to information once it was given and how it might be used were addressed through mentoring and training so that practitioner-researchers clearly explained the aims and purposes of data collection and asked for service users written permission to gather it, keep and use it. Ethics was an important part of the research training highlighting potential conflicts and issues in service providers collecting and using service users' evidence.

Analysis of the statistics generated by some instruments was also not necessarily amongst the skill sets of researchers and not transferable in the longer term when the project could no longer offer support.

> The findings from the study [using Emotional Literacy Checklist, alongside Good from Woods indicators] show the promise of woodland-based work but they are too limited by the small sample size, the specificity of the sample (young carers), and the setting to be able to make global statements about the value and effectiveness of this type of work for future participants (Acton & Carter, 2016, p. 12).

Although an aspiration of Good from Woods originally was to aggregate measured outcomes across the different cases, in practice surveys were not used in sufficient numbers to make this possible.

Some practitioners spoke of tensions between the cultural conditions they were trying to foster in order to promote wellbeing and the use of some standardised surveys. On the one hand, practitioners might want to encourage woodland-based service users to relax and engage with the here and now, without the distraction of everyday concerns. On the other, data collection required users to focus on and review their recent circumstances and health, potentially undermining the respite practitioners aimed to offer. Sometimes the administration of paperwork felt bureaucratic and too open-ended to fit with practitioners' usual role of being practically focused and helping problem solve. At worst, transfer of surveys to a woodland context could feel intrusive.

> I felt a bit embarrassed that I was giving this questionnaire out, and so many of the questions feel totally irrelevant…I don't feel any ownership over it or connection to it or its usefulness. I worry I have developed a reputation for being boring, papery and nosey and prying and concerned about how much fruit they [participants] eat each day and other 'big-brother'/naggy things!? [R2]

One solution offered to such mismatches was to ensure survey completion took place outside of the woodland activity context.

> The Emotional Literacy Checklists might be best done at home before the sets of sessions start and when they have all finished, not in the woods. [P-R]

When a good fit between tool and setting could be found, and with the right support and skills, surveys did provide valuable findings. In one case study, co-production between a school and university supported the development of a measure focused on capturing children's physical activity at Forest School. Their approach was specifically designed to suit the social forestry and primary education context and therefore overcame a number of the barriers described above.

Several projects found that surveys developed from the ground up helped them record the impacts of service provision more effectively. These had the disadvantage of not corresponding closely with similar standardised tools, being unvalidated and unfamiliar to commissioners and funders of services. However, in their favour, they sometimes better reflected the scale, setting, service users, practices, behaviours and dimensions of woodland wellbeing under research and the results could therefore be persuasive.

Frequently a way forward was a mixed-methods approach, using different approaches to triangulate the data: combining and comparing the findings of several data collection methods in order to create a balanced picture and methodology. For instance, triangulation meant that standardised survey results could be contrasted with practitioner-researcher's other findings, enriching and refining results. A weak overlap between the findings of the different methods could indicate a problem with one of the approaches (perhaps it was inappropriate to the scale and setting, service users found it difficult to respond to, service providers found it problematic to administer or it missed the data it was designed to collect, for example).

> Somewhat at odds with the lack of improvement in the Emotional Literacy scores, there was evidence of improvements in the children's well-being through their conversations and actions (Acton & Carter, 2016, p. 11).

A strong overlap of findings, several methods generating data that aligns, could indicate a robust result.

Another advantage of including a standardised tool within mixed-methods approaches was the potential to compare its findings with other interventions using the same measure, alongside its ease of interpretation by audiences familiar with it. A mixed approach also meant that no one tool dominated. This helped researchers feel they were not confined to one set of unfamiliar behaviours when researching woodland wellbeing, potentially helping them feel more comfortable in that role.

During data collection and analysis, practitioner-researchers were encouraged to explore who, what and where contributed to feeling good from woodland services (or not so good) in order to identify best practice. This was hard to achieve by standardised survey focused on outcomes.

A mixed-methods approach allowed the inclusion of tools that captured service users' journeys towards (or away from) wellbeing effects, highlighting the aspects of delivery, environment and human interaction most associated with positive (or negative) outcomes.

A Dose of Nature

The issues of employing 'standardised' evaluation and measurement in non-standardised environments also reflect wider pressures within expanded, restructured, modernised health and wellbeing services. If public health provision is to grow outside of clinical contexts, how do we conceive of and measure dosages of salutogenic interventions, such as nature, so that impacts, contraindications and side effects can be predicted?

At the time of writing, there are a proliferation of articles and projects seeking to quantify the dosage of nature that best supports human health and the most effective methods of administering it, with recommendations including 20 minutes a day based on a control study (Hunter, Gillespie, & Chen, 2019) and two hours a week from analysis of large-scale survey data (White et al., 2019). Quantification of exposure to nature is complicated, however. Studies often use geographical proximity to green space as an indicator of contact with nature. However as Silva, Rogers, and Buckley (2018) suggest, exposure is influenced by the specifics of the green space (its quality, including species richness and safety, for example) and how people interact with it. Interaction could vary in terms of: regularity and length; level of engagement (from viewing through a window to walking within it, for instance); type of interaction (with other people, alone, sporting, nature engaged, etc.); and type of contact (looking, touching, smelling, hearing, etc.) (Silva et al., 2018, p. 9547). Further, what Silva et al. (2018) refer to as the 'natural context' (time of day and year, weather conditions for instance) and the 'human context' help shape our experience of natural environments. The authors argue that variation in human contexts can include 'objective' differences such as gender identification, age, socio-economic status, exposure to nature as well as 'subjective' sociocultural factors such as personal preferences and traits and sense of connection to nature (Silva et al., 2018, p. 9548). Certainly, GfW research, discussed in the following chapters has found that variations in exposure to environmental and human contexts of woodland service provision have a strong influence on the effects of this time spent amongst trees and woodland. Subsequent changes within human contexts continue to alter the intensity of these effects years after the experience.

This variability suggests to us that currently no one standard measurement approach yet captures the diversity and specificities of a woodland context, and its human participants, for universalised quantification of the experience of woodland wellbeing. Instead, we argue that this remains a fertile area, where refining understanding of the composition of woodland wellbeing and developing measurement approaches building on that knowledge should continue to grow and evolve. There is no doubt a simple message of a dose of nature can be a useful public health guidance message to motivate more people to get out into woodland and engage with the many different effective ways to experience enhanced wellbeing. However, funders, commissioners and users of woodland-based health and happiness services should be encouraged to celebrate the diversity of delivery, responsiveness to local needs and understanding of effects that is often reflected within provision.

Challenges and advantages of standardised measurement

What can challenge use of survey instruments:

- Mismatch between impacts service delivers and what tool measures
- Cost of some measures
- Generating appropriate size sample and complete set of data
- Skills gaps (such as statistical analysis)
- Mismatch of survey questions and style of delivery with ethos and approach of service provider.

What can support use of survey instruments:

- Good match between what tool measures and audience for results
- Adapting and developing unvalidated measures to better match service provider/user needs
- Using mixed methods approaches and triangulating results

- Partnering survey tools with methods that can capture who, what and where contribute to positive outcomes
- Collaborating to fill skills gaps
- Matching survey delivery with research approaches perceived as better aligned with service provider ethos/approach and/or delivering surveys outside woodland context.

Assessing the affective can be challenging in any context as internal feelings can be lost in translation to expressed or observed emotions. The position of the researchers as members of the community being researched in Good from Woods case studies also created interesting challenges of balancing delivery with research and grounded knowledge with awareness of possible bias. Occasional clashes between methods and the material and cultural environment of woodland-based activities added another layer of complexity. Including the woodland itself within the characterisation of benefits of service delivery created a further challenge, though it was useful and revealing. However, critical engagement with these issues and collaboration with peers and more experienced researchers enabled more rounded, precise pictures of how, where and when people and the natural world coalesced in woodland wellbeing experiences than reliance on self-report surveys alone might have done. In the following chapters, we explore in more detail how different research designs unfolded in the woods and critically examine the evidence they collected.

References

Abdallah, S., Main, G., Pople, L., & Rees, G. (2014). *Ways to well-being: Exploring the links between children's activities and their subjective well-being*. The Children's Society. Retrieved from http://eprints.whiterose.ac.uk/82855/1/SCways.pdf.

Acton, J., & Carter, B. (2016). The impact of immersive outdoor activities in local woodlands on young carers emotional literacy and well-being. *Comprehensive Child and Adolescent Nursing, 39*(2), 94–106.

Bragg, R., Wood, C., Barton, J., & Pretty, J. (2013). *Measuring connection to nature in children aged 8–12: A robust methodology for the RSPB.* Retrieved from https://www.rspb.org.uk/globalassets/downloads/documents/positions/education/measuring-connection-to-nature-in-children-aged-8—12—methodology.pdf.

Carr, W., & Kemmis, S. (1986). *Becoming critical: Education, knowledge and action research.* London: RoutledgeFalmer.

Charles, L., & Ward, N. (2007). *Generating change through research: Action research and its implications* (Centre for Rural Economy Discussion Paper No. 10). Retrieved from https://www.ncl.ac.uk/media/wwwnclacuk/centreforruraleconomy/files/discussion-paper-10.pdf.

Cook, T. (2009). The purpose of mess in action research: Building rigour though a messy turn. *Educational Action Research, 17*(2), 277–291.

Gaventa, J., & Cornwall, A. (2008). Power and knowledge (Chapter 11). In H. Bradbury & P. Reason (Eds.), *The Sage handbook of action research: Participative inquiry and practice.* London: Sage.

GfW. (n.d.). *Good from Woods toolkit.* Available at https://www.plymouth.ac.uk/research/peninsula-research-in-outdoor-learning/good-from-woods/the-toolkit.

Goodenough, A. (2015). *Social cohesion and wellbeing deriving from woodland activities: Good from Woods* (A Research Report to the BIG Lottery).

Goodenough, A., & Waite, S. (2012). Book review of real world research: A resource for users of social research methods in applied settings (3rd ed.). *Journal of Education for Teaching: International Research and Pedagogy, 38*(4), 513–515.

Hunt, A., Stewart, D., Richardson, M., Hinds J., Bragg, R., White, M., & Burt, J. (2017). *Monitor of engagement with the natural environment: Developing a method to measure nature connection across the English population (adults and children)* (Natural England Commissioned Reports, No. 233). York. Retrieved from http://publications.naturalengland.org.uk/publication/5337609808642048.

Hunter, M. R., Gillespie, B. W., & Chen, S. Y.-P. (2019). Urban nature experiences reduce stress in the context of daily life based on salivary biomarkers. *Frontiers in Psychology, 10,* 722. https://doi.org/10.3389/fpsyg.2019.00722.

Kohn, E. (2013). *How forests think: Toward an anthropology beyond the human.* Berkeley: University of California Press.

McNiff, J. (2012). Travels around identity: Transforming cultures of learned colonisation. *Educational Action Research, 20*(1), 129–146.

Melson, G. F. (2013). Children and wild animals. In P. H. Kahn Jr., P. Hasbach, & J. Ruckert (Eds.), *The rediscovery of the wild* (pp. 93–118). Cambridge: MIT Press.

Myers, N. (2015). Conversations on plant sensing: Notes from the field. *Nature Culture, 3,* 35–66.

Newbury, D. (2001). Diaries and fieldnotes in the research process. *Research Issues in Art, Design and Media, 1,* 1–17.

Pavot, W., & Diener, E. (1993). Review of the satisfaction with life scale. *Psychological Assessment, 5*(2), 164–172.

Robson, C. (2011). *Real world research* (Vol. 3). Chichester: Wiley.

Silva, R. A., Rogers, K., & Buckley, T. J. (2018). Advancing environmental epidemiology to assess the beneficial influence of the natural environment on human health and wellbeing. *Environment, Science & Technology, 52,* 9545–9555.

Somekh, B., & Zeichner, K. (2009). Action research for educational reform: Remodelling action research theories and practices in local contexts. *Educational Action Research, 17*(1), 5–21.

Taggart, F., & Stewart-Brown, S. (n.d.). *A review of questionnaires designed to measure mental wellbeing.* Retrieved from https://warwick.ac.uk/fac/sci/med/research/platform/wemwbs/research/validation/frances_taggart_research.pdf. Accessed 30 June 2019.

Waite, S., Bølling, M., & Bentsen, P. (2016). Comparing apples and pears? A conceptual framework for understanding forms of outdoor learning through comparison of English Forest Schools and Danish udeskole. *Environmental Education Research, 22*(6), 868–892.

Waite, S., Rutter, O., Fowle, A., & Edwards-Jones, A. (2015). Diverse aims, challenges and opportunities for assessing outdoor learning: A critical examination of three cases from practice. *Education 3-13: International Journal of Primary, Elementary and Early Years Education, 45*(1), 51–67. http://www.tandfonline.com/doi/pdf/10.1080/03004279.2015.1042987.

Waite, S., & Waters, P. (2019). Mobilising research methods: Sensory approaches to outdoor and experiential learning research. In B. Humberstone & H. E. Prince (Eds.), *Research methods in outdoor studies.* Oxford: Routledge.

White, M., Alcock, I., Grellier, J., Wheeler, B. W., Hartig, T., Warber, S. L., … Fleming, L. E. (2019). Spending at least 120 minutes a week in nature is associated with good health and wellbeing. *Scientific Reports, 9,* 7730. Retrieved from www.nature.com/scientificreports.

5

Natural Sources of Emotional Wellbeing

What Is Emotional Wellbeing?

Emotions have not always been recognised as deeply implicated in our life chances. Instead, an emphasis on educational, cognitive and economic factors previously dominated ideas about wellbeing at population level (Acton & Carter, 2016). However, while frequently perceived as transient states, evidence suggests positive emotion can be a significant foundation for increased access to wellbeing, supporting optimistic states of mind and behaviour with longer-term beneficial impacts (Fredrickson & Joiner, 2002). Fredrickson's 'broaden and build' model (introduced in Chapter 3) predicts that in contrast to negative emotion which can limit people's capacity for action, positive emotions can alter our patterns of thought, stimulating ways of thinking that are adaptable, productive and reflective (ibid.). As discussed in Chapter 3, this type of cognition can drive exploration of new behaviours, activities and social connections, promoting personal growth and development (Fredrickson, 2013). These experimental, new

A. Goodenough and S. Waite, *Wellbeing from Woodland*,
https://doi.org/10.1007/978-3-030-32629-6_5

behaviours can 'build' our personal resources (experience, flexibility, skills, social networks, etc.) and increase our ability to achieve other forms of wellbeing, as well as further rousing positive emotion (ibid., p. 16). These 'upward spirals', driven by positive emotions, also associate them with other dimensions of health and happiness (Fredrickson & Joiner, 2002).

Recognising Emotion and Its Significance in a Natural World Context

One of the challenges in recognising emotion in research is that it is difficult to articulate and must be inferred from how we behave and what we say. It may be perceived by the individual as both embodied physiological processes and cognitive activity (Williams, 2001), but emotions also have a social dimension. It is important to note that society establishes norms for emotional conformity and that these norms influence how emotion is enacted (Denzin, 2009). Milton (2005) argues that emotions bridge the gap between nature and culture, bringing together biological (physiological and cognitive processes) and social (shaping and being shaped by environmental interactions) functions.

Emotional understanding and regulation can develop and mature (Harris, 1989). For example, while six-year-olds may find it tricky to imagine having a response other than the emotions they are currently experiencing, 10-year-olds can summon up alternative scenarios (ibid.). Children's capacity to hide negative emotions also increases as they age (ibid.). Harris (ibid.) also finds young people's emotional regulation can be affected by their sense of self-efficacy—their ability to meet their goals. Within case study research, young people in hospital can feel less confidence in the power of positive thinking to overcome adversity if they sense they lack control over their lives.

Nature and Emotion

Here, we briefly consider and recap some of the associations between emotions and the natural world (explored in more detail in Chapters 2, 3, and 9). Biophilia, nature connectedness and sociocultural attachments to nature are ways in which nature (and the woods) may support and inspire positive (and negative) emotions in people. Within biophilic theory the reasons we experience positive emotions towards the natural world seem less easily explained than origins of our phobias. One explanation of human love of the natural world is an evolved appreciation of its capacity to provide for us (ibid.). Another is that modern human's fascination with nature's variety and intricacy could be the legacy of our ancestor's successful survival strategies of interest, exploration and development of control over natural resources (ibid.). As Richardson, McEwan, Maratos, and Sheffield (2016) describe, emotion provides the spur to act, within our evolved responses to nature, in a way thought process alone could not. In fact, evolutionary theory suggests emotions developed within humans as conjoined physiological and affective adaptations helping power appropriate activity and resolution of stressful situations in our relations with natural world (ibid.).

Research suggests that one indication we are meeting our need for nature, alongside feeling relaxed and refreshed by green spaces, can be experiences of positive emotions in relation to the natural world (Lumber, Richardson, & Sheffield, 2017). These could include feeling good when engaging the senses with nature and feeling love for nature following such engagement, as well as experiencing compassion and concern towards the natural world that may support future caring attitudes and actions (ibid.).

Our cultural attachment to nature as part of our heritage and biography (embedded within religious, national, local narratives of who we are) may engender positive emotions when we spend time in natural landscapes. As UK studies of public views of woodland attest, adults are aware of trees' historical and cultural significance, as symbols of nature, life, the state of the environment and British values (O'Brien, 2005; Carter, O'Brien, & Morris, 2011). We can also have personal, emotional attachments and aversions to particular natural places, spaces, species and landscapes, reflecting our background and experiences (Chawla, 1990). As Chawla (1999) reflects, our personal narratives of who we are often include memories of natural settings, and the emotional significance of the time we spent in them is constantly revisited, informing our current sense of self.

Last, but not least, the type and length of activities, people and cultural values we may engage with while spending time in nature could all act as prompts to us feeling positive or negative moods and feelings. In practice, it seems that the range of ways in which nature might foster the conditions for positive emotion or directly inspire it can be difficult for people to untangle from the impacts of contemporaneous sociocultural activity.

For instance, Maller (2009) interviewed staff in schools and environmental education organisations in Australia to determine the potential of hands-on contact with nature to improve emotional wellbeing and social relationships in 'normal' primary-aged school children. While organisation staff reported that unstructured individual self-directed activities were more closely linked to children's mental, emotional and social wellbeing, organisation managers placed greater emphasis on adult-led, structured activities. In this instance, the entanglement of cultural and natural impacts on emotional wellbeing was interpreted differently by each group. School staff and head teachers on the other hand tended to see all time spent in nature as beneficial for children.

Barton, Hine, and Pretty (2009), studying the effect of walks in National Trust sites, found emotional wellbeing was enhanced through both exposure to nature and participation in exercise, with feelings of anger, depression, tension and confusion reduced and vigour increased. While identifying that activity and environment were both factors in effects on positive emotion, they did not analyse correspondence between the material landscapes walked within (woodland, fenland, heath or river valley) and types of response. Marselle, Irvine, Lorenzo-Arribas, and Warber (2015) suggest in fact that the extent to which natural landscapes are subjectively perceived to be 'restorative' (explored in Chapter 2) is a key factor in their positive effect on emotion and their reduction of negative emotion after activity. Their case study explored the UK walk for health scheme (one of the largest UK natural health interventions) with participants aged 55+. Environments that reinforced a sense of 'being away'—psychologically and physically distant from the usual demands on attention—and provided easy fascination with a natural landscape, which walkers also feel at one with, were perceived as most restorative. Some short-term emotional wellbeing benefits of nature may arise through individual's perceptions of that environment as restorative, rather than directly from specific features such as diversity of flora or fauna.

Lee et al. (2017) in their systematic review of forest therapy (visiting a forest and/or engaging in specific therapeutic activities amongst trees) programmes designed to decrease levels of depression amongst adults found positive effects for people with identified health problems. However, like Barton et al. (2009), they found a 'ceiling effect' in evidencing change for apparently healthy adults (at a certain level of good health, forest interventions were no longer increasing health levels) (Lee et al., 2017). It may be that those who already feel quite well tend to seek out such places and/or activities that can support them to feel well, in a virtuous circle of reinforcement.

This idea of positive reinforcement is partially supported by Finnish research (Korpela, Borodulin, Neuvonen, Paronen, & Tyrväinen, 2014) exploring whether a relationship between time spent in nature-based recreation and emotional wellbeing can be accounted for or altered by multiple factors associated with activity in nature. The variables explored for their effect on this relationship were restorative experiences, social company or duration of people's (most recent) visit in nature. 90% of the respondents reported forest as the context for their last nature-based activity. The study found a positive relationship between the self-reported time spent in nature-based recreation and subjective emotional wellbeing. Restorative experiences in the most recent visit, but not company or length of visit, were positively implicated in this relationship. Feeling calm, revitalised, 'forgetting everyday worries', gaining clarity over one's thoughts and optimistic about the future were important factors in associating emotional wellbeing with nature visits (Korpela et al., 2014, p. 5).

Clearly, relationships between the individual, natural and cultural contexts intersect in complex ways to create conditions for emotional wellbeing, and these favourable conditions are subject to change over time. However, nature's restorative and stress-reducing effects, direct or perceived, including a sense of getting away from everyday concerns and contexts, are an important factor in enabling positive emotions following time spent in nature. The striking material difference of woods, in terms of their rich biodiversity and our perceptions of their 'naturalness' or 'restorative' value, in comparison with the places we normally reside, may impact positively on our wellbeing. In a sense, such natural places may be regarded as partners in managing emotions (Goodenough, Waite, & Wright, under review), just as people can help to support our emotional health.

This chapter focuses on two Good from Woods studies that captured evidence of people feeling good about time spent in woodland. It engages with the challenges of identifying how feeling good is associated with the material environment, people or activity and proposes some of the factors that can best account for this report in order to contribute to better understanding of how woodland and activities within it may contribute to emotional wellbeing. Both case studies were located at Embercombe. Emotional wellbeing was identified in the GfW framework as indicated when people were experiencing positive emotions and moods and an absence of negative emotions and moods. Indicators of positive emotional wellbeing included feeling even-tempered, relaxed and feeling optimistic about the future.

The Woodland Site

Embercombe is a charity and social enterprise on a 50-acre site in Devon which at the time of the research focused on land-based education. Eight permanent members of staff, some freelance support and a changing community of volunteers together ran various activities. These included residential and day visits for schools, working with 'disadvantaged' teenagers, corporate training, volunteer and apprenticeship programmes.

The wooded area is their privately owned hundred-year-old broadleaf woodland of medium density. It is a mixture of oak, hazel, birch, ash and hawthorn trees of different ages and sizes, varying between massive old oaks, slim towering birches, hazel coppice and more recently planted saplings. The ground is carpeted with bracken, ferns, brambles and, in April/May, vibrant bluebells. The woodland occupies the side of a valley opposite the residential and meeting areas. The ground slopes quite steeply and tends to get slippery and muddy. Jessie Watson-Brown, the

Fig. 5.1 Looking towards Embercombe's woods (*Source* Goodenough, 2015)

practitioner-researcher at Embercombe, carried out two pieces of Good from Woods research in these woods (Fig. 5.1).

Case Study 1: Reconfiguring Relationships for Emotional Wellbeing

The first of Jessie's case studies explored here aimed to find out what factors in the woodland-based experiences of Steiner School students on a residential visit were associated with wellbeing outcomes. She was interested in whether what young people felt to be significant was broadly the same as staff expectations. Were Embercombe programmes achieving the outcomes they intended?

The residential visit studied was a ten-day, one-off programme. There were 25 participants (12 female, 13 male) aged between 15 and 16 years from Hereford Steiner School. About 40% of activities took place in the woodland, and as with the first case study, this formed a point of comparison, helping students describe what they got out of woodland-based activity as opposed to activity based elsewhere. These were mostly purposeful activities such as firewood collecting, tree planting, cooking, gardening and building, but they also engaged in creative activities and team-building exercises.

Staff at Embercombe and the school sending students to the residential anticipated that the ten-day visit would result in several effects on young people that align well with the outcomes framework for 'natural learning' (outside the classroom), proposed by Malone and Waite (2016). Staff intentions for woodland activity impacts had aspects of all five of the authors' (2016, p. 5) suggested outcome areas: 'a happy and healthy mind', 'a sociable and confident person', 'a self-directed learner', 'an effective contributor' and 'an active global citizen'. Staff widely expected young participants to experience positive emotions and moods in the woodland context, anticipating that its restorative influence might both relax and invigorate students.

> Woodland in my experience makes one feel well and makes one feel happy – not a joyous jumping up and down happy but a contented happy – I think they have a very peaceful – a combination of peaceful, calming influence, but there's something quite alive about a woodland' [S2H].

> Something about [Embercombe's woodland based activity] enhancing people's hopes and aliveness with life [S3HE].

> Very energizing, people seem to be very, you know, motivated in that sort of [woodland activity] environment [S4E].

They also strongly felt that woodland activity could have an impact upon young people's social relations, with chances for them to support others and experience similar support themselves.

> I'm expecting a huge shift in the social dynamic of this class while we are down there pushing them in groups to challenge them with, you know they are not going to be with their friends. I mean in their little social groups and I can't say whether it's just the woods or the entire experience [S4E].

> Experiencing themselves and each other in a whole different way being out of doors and you know maybe experiencing their friendships among their contemporaries in a whole different way [S3HE].

Alongside these outcomes, staff hoped young people might feel engaged in a relationship with nature during activities and at one with their environment. Finally, they imagined that young people could experience psychological wellbeing, feeling purposeful and competent during woodland-based work and developing aspects of themselves.

The mechanisms for achieving these outcomes included simply spending time in woodland. However, some identified 'doing activities' and 'working' as the main contributory factors, while other stakeholders focused on the social interactions of the group, mentioning the peer pressures often experienced by this age group, and their expectation that the woodland activities and residential experience as a whole would have a significant impact on participants in this respect.

Over the course of the ten-day residential, Jessie used a mixed methods approach to explore whether staff assumptions were played out in the experience of the young people. A paper-based questionnaire exploring health and wellbeing was used that students could complete on arrival at Embercombe and just before leaving. Despite Jessie's request for them to fill it in without conferring, students tended to work with peers to come up with answers. Jessie also found herself feeling self-conscious trying to encourage students to engage seriously with this exercise and was concerned she might be perceived as 'boring', 'papery' and 'prying' [Jessie PR]. As the ten-days went on, she experimented with group discussion and found initially that this only suited some of the more confident students.

> quieter participants were able to avoid saying very much and not be heard, even when I facilitated the discussion. I think they were saying what they think I want to hear. The participants often tended to talk about facts rather than feelings [Jessie PR].

Jessie planned individual interviews to fit around woodland activities, asking students if they had a few minutes to chat around their practical or learning tasks and asking them some prepared, but fairly open questions.

Again, she found young people seemed guarded and low key in their responses.

> very tricky, [participant] self-conscious and listened to by [friend] and I noticed [friend] was smirking at the questions I was asking (or at the fact that his peer had to answer them). [Jessie PR]

Halfway through the residential, Jessie began to get a sense that things were shifting. Young people were starting to feel more able to open up, and she now felt less at odds in her role as researcher. She reflected that young people's self-conscious engagement with the research mirrored the intense social pressures they were experiencing amongst their peers and anticipated by some staff. Her own feelings of embarrassment and concerns of being perceived as 'boring' seemed in part due to her temporary involvement in this intense and self-aware social atmosphere. Some of the shift towards students sharing more may have been due to increased trust building between them and Jessie. Certainly, young people were beginning to appreciate interactions with other adult staff at Embercombe. Equally, they were also starting to feel safer and more supported within their peer group.

> The biggest thing was the people – so friendly [3F].

> The other reason I really liked the logs was because [staff member] was so funny which made the work easier [Map—anonymous].

> I have got to know and trust the classmates more [3M].

However, Jessie also linked it to a change in young people's perceptions. Residential pupils were beginning to experience and be more aware of the woodlands as a place of emotional change and were therefore better able to understand Jessie's questions about enjoyment, challenge and wellbeing as relevant.

> To begin with, there was little sharing of feelings, whereas the end of the programme saw participants openly express themselves, particularly in their final presentations [on their experience at Embercombe] [Jessie PR].

Jessie experimented with asking students to annotate maps of the site with their experiences and responses to them, a useful method that afforded them privacy and anonymity in reporting their individual views. However, as the residential drew to a close, several students were beginning to share with Jessie and the staff how moments of affect in the woods at Embercombe were affecting them. These included both positive and negative feelings and moods

> Initially, [moving wood] seemed pointless. After, now, I see the point, see the bigger picture. You feel you can affect something [3F].

> Didn't enjoy log heaving we did; the activity was too physical for me to endure. I got very worn out (Map—anonymous).

> looking back and seeing how much you've done. You look back and think "my God, we've moved ten times my body weight in wood". Really satisfying [2F].

> Didn't like wood carrying...Couldn't do the activity, liked the leader, liked the woods [12M].

Woods were also referenced in a way that acknowledged their capacity to be, restorative, relaxing, calming and reflective spaces, in the way that staff had anticipated, though this was not a universal experience.

> It's got like a peaceful kind of feel to it though hasn't it? Relaxing and calm sitting in the woods [1F].

> [The woods] Can be relaxing, but can be uncomfortable –[I] prefer open spaces...[where I] feel free to go anywhere, in a forest there are set paths and brambles in the way you have to pick your way round things [6M].

> I went to the woods to think and be with myself, to think about...find out who I was, run around and kicking a few trees, then I got back and wrote loads down, sitting in a ditch by a stream [10M].

As the days wore on, the woods appeared to become more socially and culturally relevant spaces to students—more worthy of both spending time in and discussing. Reflective links were made between feeling good in the woods in the present and experiences of wellbeing in outdoor settings in the past and even feeling good in the future.

> I connected to my old outdoorsy self when I was making the face on the tree – I feel so much like my old self and that's significant [2F].

> Have you ever been to the Forest of Dean? Its near where I grew up…It's nice to be in the woods, to explore, wander around. I feel calm and lively, different things going on, bugs and birds all around [8F].

> I used to be outside loads, now I just watch TV. I want to go back to my old ways of being outside more [11F].

Top 3 messages from Steiner School residential case study

1. Time spent in a woodland environment by young people can provide access to positive emotions in the present, potentially encouraging a reinforcement of individual's capacity to access wellbeing from natural environments.
2. In highly pressured peer environments, positive relationships with adults can support young people to feel safe and supported.
3. For difficult to engage young people, building research into the activity may be an effective research tool.

Case Study 2: Escaping the Everyday for Emotional Wellbeing

Jessie's second case study explored a year-long, three days a month, programme—the 'Young Leaders Project' (YLP). During their regular stays, seven young people (aged between 17 and 21, 3 female, 4 male) stayed on site and carried out practical tasks such as tree management, entailing hard physical labour, during their regular 3-day stays. Jessie again aimed to discover whether young people reported similar experiences and wellbeing outcomes to those anticipated by the staff. Information about any mismatches would be used to refine and improve planning and practice for future groups.

Students entered the YLP programme through referral from a regional youth organisation supporting them during their time in care. About 15% of activities were woodland based, including coppicing, coppice timber processing and tree planting. Other YLP activity included cooking, gardening, building, creative and team-building exercises.

Staff expectations for outcomes of YLP were similar to those for the Steiner School residential, but with increased emphasis on nature engagement and psychological outcomes of feeling competent. In this case, purposeful action for nature was at the forefront of anticipated outcomes, with social and emotional wellbeing secondary (Malone & Waite, 2016). Staff believed that spending time in the woods would result in these types of wellbeing, but some focused on the opportunity for having 'time-out' and just 'being', two emphasised 'doing activities' and 'working' as contributory. Two social workers who worked with the young people at home anticipated that simply 'being away' from their daily life would have a positive impact on the participants.

Jessie found that audio-recorded individual conversations during activities worked well as participants were free to talk with less worry of peer pressure and less disruption to their planned programme of activities. However, this being a self-contained small group working together over a period of time and familiar with Embercombe's staff (including Jessie) and setting, she found audio-recorded group interviews were also successful and beneficially informal. Quieter group participants' opinions were sometimes overshadowed by more dominant participants but having a familiar member of staff facilitate ensured everyone could contribute.

> I know them quite well and feel they would open up to me…just because I've spent time with them and helped them on things…having a good relationship with them will be beneficial to this [research] so that's interesting

> to notice and probably will actually in a sense improve their experience of their time here, because I'm making more of an effort to really engage with them and listen to them [Jessie PR].

As with the school residential case study, all the data was reviewed and coded within the GfW wellbeing framework and indicators described in Chapter 3. This helped identify kinds of wellbeing being accessed in the woods and who, what and where led to these experiences.

The most common wellbeing outcome reported by YLP participants was feeling relaxed. This positive emotional effect was attributed by young people to spending time away from their daily lives and worries, the atmosphere of the woodland and working in the woods. An absence of negative emotions and moods was the second most frequently reported effect which was also associated with being outdoors in woodland and again, away from normal stresses.

> it was peaceful [in the wood], I don't have half the stresses I normally put on myself unnecessarily, cos it wasn't there to remind me, which was nice. So, I didn't worry and go off on things and get a grump on cos things weren't going my way [gp2].

> I've got a load of physical energy, but it's just calm, it's just like...the wood, the woods in general just makes me calm [Mp3].

> work in the woods, overall, even though its hard work, it's enjoyable and I think that's why I find it relaxing cos I enjoy doing it [Dp1].

> For some reason whenever I go into any woodland sort of space, I've got this constant song in my head and it literally says what it means. You know that song off Lion King "no worries", [...] it just makes me forget about everything [Jp2].

> normally when I'm in the woods I just let my mind wander. But when I'm normally at home it normally wanders about negative things really like how's everything going to fit together. But when I'm in the woods I just let it go and occasionally have a laugh and a joke...when I'm doing stuff I enjoy in the woods it just seems to go to positive things which makes me

> enjoy the experience more cos then I'm just thinking constantly about the good times, good things, which is always good [Cp2].

Positive emotions and mood were the fourth most common outcome of woodland activity and linked to a combination of collaborating in purposeful work and the woodland context, where they were learning new skills within a developing peer group. Enjoying and being confident in physical activity were the third most frequent effect of time spent in woodland, also linked to positive emotions experienced when working hard and contrasted to their 'normal' lives.

> my body felt confused. In the way that I'm not being used to being outside for such a long time [Cp1].

> It's a good kind of ache, where you feel like you've achieved something instead of sitting around at home doing nothing [P7].

Sometimes enjoyment developed in hindsight, with senses of achievement in the work.

> we're always put down on stuff we're not doing good…instead of being told all the time you need to stop doing this and that… it's good that someone can turn round and says oh that's good you've planted something that will last for hundreds of years, done something that will help future generations. It sort of like makes you feel better about yourself [C].

Some young people were able to clearly articulate the role the natural world had in making them feel positive emotion; others expressed how they valued connection with the woods. Feeling engaged in a relationship with nature was Jessie's fifth most common finding, with its soft fascination.

> Like cities and stuff like that, because you've just got everything that's a distraction, but here, there's not much distraction. But in a way there is, there's lots of distraction. Because you've got all the trees, but they're [trees] not as like simple as you know like buildings. It's really nice, I like it (MP3).

Top 3 messages from Embercombe's Young Leaders case study

1. A change in context (people, place and activity) and getting away from daily lives was important in facilitating experiences of wellbeing, especially for these young people from disadvantaged backgrounds.
2. Performing meaningful work-based tasks combined with experiences of praise and appreciation from staff provided the young people with a sense of competence and purposefulness. Again, this seemed different to their usual experience.
3. Woodland allowed different relationships and feelings to emerge.

The restorative difference of woodland environment and context helped students at Embercombe, feel relaxed, emphasised a sense of getting 'away' from the concerns and trappings of everyday lives and able to enjoy activity or social relations. These effects proved important within research case study 2 with young people in care. Some experienced woodlands to be restorative, enjoying their diversity, complexity and the profound sense they provided of 'getting away' (Marselle et al., 2015; Waite, 2013). Woodland supported experiences of getting away from everyday concerns and behaviours, reinforced by new social bonds and connections, and potentially enabled feeling and behaving differently.

What was less clear was how resilient these happy moments were. For some, heading back to their day-to-day settings and activities were reported to cancel benefits out quickly. The extreme contrast between the woods and everyday lives proved a barrier to experiencing ongoing wellbeing: restricting the experience to an intense opportunity for respite. When YLP students were asked about their experiences on the programme six months later, most wellbeing experiences endured in their memory, how being away from worries and daily lives helped them feel relaxed and positive, and the change in place had supported feelings of competency, purpose and being accepted were particularly recalled. However, connection to the natural world was not mentioned in follow-up interviews. This longitudinal finding was not consistent across GfW case studies; indeed in several instances, a sense of relationship to nature remained one of the strongest memories of woodland activity. However, in this instance, young people returning to their home environment also returned to their 'habitus', dispositions or styles of behaviour and emotional reactions that are associated with that social and cultural setting (Bourdieu, 1977). The weight or density of these cultural expectations and patterns may have shut down newer, less ingrained responses associated with the woods (Waite, 2013). The resource of calm moods and feelings inspired within woodlands during YLP seemed strongly tied to that restorative context, yet six months later young people did not recall nature as one of its sources.

These findings raise questions as to whether accessing wellbeing via nature-based interventions can be significant if it is limited on return to less natural and novel contexts. How might woodland inspired emotional wellbeing be made more resilient in the face of structural disadvantage or rather perhaps, how might we change the circumstances in those disadvantageous contexts so that emotional wellbeing can be maintained? Harris's (1989) finding that young people can feel less hopeful about the power of positive thinking to overcome adversity if they feel a lack of control over their lives may be relevant here. Did the positive emotions and associated behaviours seem less plausible in difficult to control home circumstances? Likewise, YLP students no longer experienced the restorative experiences in nature back within their everyday contexts. Might interventions that rely on nature's calming and refreshing effects need to help participants find similar settings closer to home? Marselle et al.'s (2015) finding that the perception of restorativeness can be more influential than specific natural landscape features suggests that nearby nature might serve a similar function of feeling 'away'.

Contexts for Emotional Wellbeing

Notable in both studies is that many people felt good because they felt different. This sense of difference was linked at times with the relative novelty of spending time in woods, woodland activities or the temporary social communities formed around doing things in the woods. Doing something new has frequently been associated in research with positive wellbeing and 'neophilia' or novelty-seeking behaviours have been cited as important predictors of life satisfaction (psychological or eudaemonic wellbeing). Woods provide important contexts for novel experiences in the context of our predominantly urban lifestyles and spaces. However, once Embercombe's woods became more familiar, simply spending time in nature was associated with emotional wellbeing for some, without them necessarily doing anything or seeking out the new.

This echoes other studies that suggest familiar places can also bring a heightened sense of wellbeing (Bell, Phoenix, Lovell, & Wheeler, 2015; McCree, Cutting, & Sherwin, 2018) through the laying down of layers of memory and association with good experiences (Waite, 2007). In residential case study, some young people connected positive memories of time spent in nature from their past with their growing familiarity and enjoyment of the woods at Embercombe and a sense that these behaviours, moods and feelings might be useful ways of being again in the future. This required both that young people had affective memories of natural landscapes that could be linked with new experiences and that they believed such opportunities might be open to them again. Denzin (2009, p. 58) argues 'Emotionality is a circular process' where past and future meet in the present. While it is only possible to be in one emotional situation, 'the emotional consciousness and emotional behaviour I am currently occupying' (ibid., p. 79), the self brings different times and situations together in that one moment. Interventions aimed at increasing emotional wellbeing need to be conscious of how positive emotions associated with woodland activities map onto past experience and find potential to be continued into the future.

Perhaps most sustainably, the benefits of woodland emotional wellbeing might be maintained by creating more treed urban environments and promoting more regular access to these and rural woods for young people, while challenging some of the social norms that mean green areas are less commonly used by young people in their daily lives (Natural England, 2018) However, it's important to recognise that for some young people, working in woodland and making a difference (also see Chapter 7) were an important source of relaxing and calm (emotional wellbeing), as well quieter and passive enjoyment of woodland.

Implications for Promotion of Emotional Wellbeing: 'Feel Good Factors'

Young people (case study 1) arriving at a ten-day residential stay at a land-based education centre could not articulate what about woods might be of benefit to them. Discussion of woodland environments and activities felt of little relevance to them and the researcher collecting data felt like an irritating outsider in trying to generate this data. Subsequent reflection on these feelings suggested that it was the extreme peer pressure experienced amongst the group to conform to group norms that presented barriers to peer and intergenerational social connection and sharing of ideas. This was especially evident in the initial stages of the research, and to some extent, it is likely to apply when assessing short-term interventions, where there is limited time to build up relationships of trust.

However, by the end of their stay, young people identified the woodland as a key context in which they had been able to connect socially with classmates; they also mentioned valued support from adults. The woods themselves seemed to be a vital resource for calming and relaxing and operated as such without 'activity' or 'social' mediation at times. In addition, the wood was perceived as a space in which they could reconnect with memories, including embodied memories of being younger and more engaged with outdoor spaces, recollections that helped them feel good and access benefits from that environment again. This study highlights the impact of contemporaneous sociocultural contexts and behavioural norms upon our ability to access emotional wellbeing from natural environments. Once 'memories' of childhood and past outdoor activity were reignited, young people reported feeling more assured of the value of time spent in woodland and secure in the positive feelings engendered there.

This points to two potentially successful strategies towards promoting emotional wellbeing outcomes:

- Memory prompts in creating a narrative connecting past, present and potentially future experiences of emotional wellbeing—accelerating the benefits from shorter term interventions and
- Establishing an emotionally 'safe' trustworthy environment to enable emotional responses to be stimulated or shared.

In case study 2, as staff had anticipated, being away from the usual stresses of their urban lives was a key factor in promoting the young people's emotional wellbeing. This was initially attributed to the novelty factor but given the repetition of monthly visits over an entire year, it seems that familiarity may also have brought deeper emotional restoration and calming of sensitivities. Coupled with increases in psychological wellbeing through a growing sense of achievement and satisfaction, the outlook for the young leaders was more hopeful. Although the research did not follow up any long-term changes to their emotional responses in their home environment, there appeared to be a development of more positive attitudes over time. As Waite (2013) suggests, both outdoor contexts and novelty may separately and in combination contribute to a less rigid cultural density of established social norms, making alternative ways of being and feeling possible for people.

This indicates potential for emotional wellbeing through:

- Ensuring autonomous opportunities to express different emotions and try out new behaviours in outdoor activities,
- Allowing time for positive changes to become embedded as alternative ways of being in other settings
- Supporting sustainable wellbeing through linking to and promoting access to woodland in home settings.

Capturing Emotional Wellbeing from Woodland Experiences

For those practitioners and researchers wishing to explore the role of emotions in fostering woodland wellbeing, data collection and interpretation of emotional wellbeing might take the following basic steps:

- Uncover initial emotional understandings, motives, intentions and feelings,
- (Re)Present this in thick descriptions,
- Observe embodied demonstrations of self-feeling,

- Consider any past influences,
- Capture reflections and meanings emergent during and after the experience,
- Contextualise with social and cultural contexts that have shaped individual subjective experience.

Adapted from Denzin (2009, p. 255).

References

Acton, J., & Carter, B. (2016). The impact of immersive outdoor activities in local woodlands on young carers emotional literacy and well-being. *Comprehensive Child and Adolescent Nursing, 39*(2), 94–106.

Barton, J., Hine, R., & Pretty, J. (2009). The health benefits of walking in greenspaces of high natural and heritage value. *Journal of Integrative Environmental Sciences, 6*(4), 261–278.

Bell, S. L., Phoenix, C., Lovell, R., & Wheeler, B. W. (2015). Seeking everyday wellbeing: The coast as a therapeutic landscape. *Social Science and Medicine, 142,* 56–67.

Bourdieu, P. (1977). *Outline of a theory of practice.* Cambridge: Cambridge University Press.

Carter, C., O'Brien, L., & Morris, J. (2011). *Enabling positive change: Evaluation of the Neroche landscape partnership scheme.* Farnham: Forest Research.

Chawla, L. (1990). Ecstatic places. *Children's Environments Quarterly, 7,* 18–23.

Chawla, L. (1999). Life paths into effective environmental action. *The Journal of Environmental Education, 31*(1), 15–26.

Denzin, N. (2009). *On understanding emotion.* Brunswick, NJ: Transaction Publishers.

Fredrickson, B. L. (2013). Positive emotions broaden and build. In P. Devine & A. Plant (Eds.), *Advances in experimental social psychology* (Vol. 47, pp. 1–53). Burlington: Academic Press.

Fredrickson, B. L., & Joiner, T. (2002). Positive emotions trigger upward spirals toward emotional well-being. *Psychological Science, 13*(2), 172–175.

Goodenough, A. (2015). *Social cohesion and wellbeing deriving from woodland activities: Good from Woods* (A Research Report to the BIG Lottery).

Goodenough, A., Waite, S., & Wright, N. (under review). Place as partner: Material and affective intra-play between young people and trees, *Children's Geographies*.
Harris, P. (1989). *Children and emotions*. Oxford: Blackwell Publishing.
Korpela, K., Borodulin, K., Neuvonen, M., Paronen, O., & Tyrväinen, L. (2014). Analyzing the mediators between nature-based outdoor recreation and emotional well-being. *Journal of Environmental Psychology, 37,* 1–7.
Lee, I., Choi, H., Bang, K.-S., Kim, S., Song, M., & Lee, B. (2017). Effects of forest therapy on depressive symptoms among adults: A systematic review. *International Journal of Environmental Research and Public Health, 14,* 321.
Lumber, R., Richardson, M., & Sheffield, D. (2017). Beyond knowing nature: Contact, emotion, compassion, meaning, and beauty are pathways to nature connection. *PLoS One, 12*(5). Retrieved from https://journals.plos.org/plosone/article?id=10.1371/journal.pone.0177186#sec001.
Maller, C. J. (2009). Promoting children's mental, emotional and social health through contact with nature: A model. *Health Education, 109*(6), 522–543.
Malone, K., & Waite, S. (2016). *Student outcomes and natural schooling: Pathways form evidence to impact report 2016*. Retrieved from https://www.researchgate.net/publication/305790449_STUDENT_OUTCOMES_AND_NATURAL_SCHOOLING_PATHWAYS_FROM_EVIDENCE_TO_IMPACT_REPORT_2016.
Marselle, M. R., Irvine, K. N., Lorenzo-Arribas, A., & Warber, S. L. (2015). Moving beyond green: Exploring the relationship of environment type and indicators of perceived environmental quality on emotional well-being following group walks. *International Journal of Environmental Research and Public Health, 12,* 106–130. https://doi.org/10.3390/ijerph120100106.
McCree, M., Cutting, R., & Sherwin, D. (2018). The hare and the tortoise go to Forest School: Taking the scenic route to academic attainment via emotional wellbeing outdoors. *Early Child Development and Care, 188*(7), 980–996.
Milton, K. (2005). Emotion (or life, the universe, everything). *The Australian Journal of Anthropology, 16* (2), 198–211.
Natural England. (2018). Monitor of engagement with the natural environment: National survey on people and the natural environment (Headline Report). Retrieved from https://www.gov.uk/government/statistics/monitor-of-engagement-with-the-natural-environment-headline-reports-and-technical-reports-2016-2017-to-2017-2018.
O'Brien, L. (2005). Publics and woodlands: Well-being, local identity, social learning, conflict and management. *Forestry, 78,* 321–336.

Richardson, M., McEwan, K., Maratos, F., & Sheffield, D. (2016). Joy and calm: How an evolutionary functional model of affect regulation informs positive emotions in nature. *Evolutionary Psychological Science, 2*(4), 308–320. Retrieved from https://link.springer.com/article/10.1007/s40806-016-0065-5.

Waite, S. (2007). 'Memories are made of this': Some reflections on outdoor learning and recall. *Education 3-13; International Journal of Primary, Elementary and Early Years Education, 35*(4), 333–347.

Waite, S. (2013). 'Knowing your place in the world': How place and culture support and obstruct educational aims. *Cambridge Journal of Education, 43*(4), 413–434.

Waite, S., & Goodenough, A. (2018). What is different about Forest School? Creating a space for an alternative pedagogy in England. *Journal of Outdoor and Environmental Education, 21*(1), 25–44.

Williams, S. (2001). *Emotion and social theory*. London: Sage.

6

Natural Sources of Social Wellbeing

What Is Social Wellbeing?

Inevitably, as we have seen in Chapter 5 on emotional wellbeing, the dimensions of wellbeing overlap to some extent. Social wellbeing is particularly challenging to contain in a short definition and set of indicators. It can refer to our sense of ourselves within our interpersonal relationships. However, it can also describe social aspects of other types of wellbeing—forms of feeling good that are hard to achieve without social support and a sense of social safety. As a reminder, social wellbeing was initially identified in our project as when people in woodland felt confident, accepted, safe and supported *within and through* social relationships, supporting others through social relationships.

Social Animals

We are by nature social animals according to Aristotle. It is therefore unsurprising that human flourishing is highly dependent on social wellbeing.

A. Goodenough and S. Waite, *Wellbeing from Woodland*,
https://doi.org/10.1007/978-3-030-32629-6_6

This has not always been acknowledged in macro definitions of wellbeing emphasising economic factors, however. As described in Chapter 3, research has established that in modern, wealthy individualistic societies where traditional collective ties and associations are less influential, interpersonal relationships and social connections are valuable sources of wellbeing (Helliwell & Putnam, 2004). In fact, it has been argued that, within relatively wealthy economies, levels of SWB can best be predicated by establishing the breadth and strength of an individual's social networks (ibid.).

Social wellbeing, however, can be understood in ways that go beyond capturing our sense of ourselves within our social relationships. Frazer (2017), for example, poses a broader question about the social dimensions of wellbeing, asking 'can you be happy when others aren't?'. Drawing attention to a relational dimension to subjective wellbeing (SWB), Frazer suggests that personal wellbeing must take account of others in society, arguing 'social responsibility, equity and reciprocity are fundamental qualities in any meaningful formulation of wellbeing'. Teghe and Rendell (2005: 2) citing Keyes, (1998), set out five social dimensions to wellbeing: 'social acceptance (accepting others as they are), social actualisation (positive comfort level with society), social contribution (a feeling that one has a contribution to make to society), social coherence (understanding the social world as predictable comprehensible) and social integration (feeling as a part of the community)'.

These wider understandings of wellbeing as social and relational can be coupled with ideas discussed Chapter 3, that wellbeing does not reside as a stable trait inside a person but shifts in relation to societal and cultural values, places, people and time (Adams, 2012). It is a composite of individual, societal, economic and ecological factors: not a 'series of changes brought about in static states by the reactions of dependent variables to independent variables, but an ongoing process always embedded in interdependent relations' (Kasper, 2009: 4).

Yet, current governmental interest in social wellbeing still seems driven largely by economic factors and the opportunity to forestall later societal problems by upstream interventions. For example, NICE (UK's National Institute for Health and Care Excellence, 2019) guidelines regarding children and young people of secondary school age define social wellbeing as 'good relationships with others', adding that means having 'no behavioural problems such as disruptive, bullying or violent behaviour'. For older people, poor social wellbeing is being increasingly linked with social isolation and loneliness. In both cases, avoiding negative outcomes rather than providing positive social impetus seems to motivate. Notwithstanding the motivation, preventive public health interventions such as nature-based interventions and encouraging volunteering are to be welcomed as there is ample evidence that these do make a positive difference to people's social wellbeing (Maller, Townsend, Pryor, Brown, & Leger, 2006; Stigsdotter et al., 2011) and link with psychological wellbeing of purposefulness.

Shared Social Encounters Within Nature

Natural spaces appear to be environments supportive of human social interaction. For instance, they provide a meeting place and setting for social interaction and integration. In the UK, they are also frequently the venue for activities supportive of social connection and wellbeing. At the same time, as discussed in Chapter 2, they provide the engaging and restorative environment that through improving health and wellbeing can benefit interpersonal and social interaction.

Natural spaces can provide an environment where people can see, bump into and spend recreational time with one another. Shared experiences with other people (O'Brien & Morris, 2014), spending time with friends in a forest physical activity programme and visiting peri-urban forests (O'Brien, Townsend, & Ebden, 2008) have all been found to be beneficial. Case study research showed that more time spent in forests and parks improved self-satisfaction and social contacts amongst Iranian adolescents, for instance, especially for older young people and boys (Dadvand et al., 2019). However, fears around safety, particularly for women, can affect recreational access to woodlands (Burgess, 2002; Forest Europe, 2019: 31; Morris, O'Brien, Ambrose-Oji, Lawrence, & Carter, 2011).

Taking part in interventions focused on and taking place within natural settings has demonstrated specific effects on social relations, alongside personal wellbeing. Research for the Wildlife Trusts by Rogerson, Barton,

Bragg and Pretty (2017) concluded that volunteering in natural environments not only resulted in changes in the way individuals felt about themselves but was associated with change how they felt about other people. Volunteers, for example, viewed others more positively, took part in affirmative relationships with new sections of society and reduced their sense of social isolation. The social value of spending time with others in natural space focused interventions has also been noted in a recent systematic review about places, spaces, people and wellbeing (What Works Wellbeing, 2018). This appraisal similarly found green and blue space interventions offer opportunities to participate in activities that improve social interactions and increase community integration by mixing generations, as well as cultural and socio-economic groups. For example, creating an informative and accessible trail together, through woodland, contributed to solidarity and tolerance between participants, benefiting collective social bonds (Vering, 2006). Improving green and blue spaces can also enhance interpersonal relationships by providing something for families to do together (What Works Wellbeing, 2018). Family wellbeing and interactions might also be benefited through parents feeling they have met their children's need to be in contact with nature and witnessing them benefit from its engaging and restorative qualities (Goodenough, Waite, & Bartlett, 2015). The What Works Wellbeing review (2018: 8) suggests that green/blue space interventions can improve social networks, interactions, capital and cohesion.

Biophilic and Social Benefits of Time in Natural Outdoor Spaces

Many studies have shown that social interaction and relation is often mentioned in association with enjoyment of nature and natural connection (for example, Cameron-Faulkner, Melville, & Gattis, 2018), and disentangling the effects of each can be a challenge. Cartwright, White, and Clitheroe (2018) point out that several studies imply that connectedness in one domain (social or biophilic) may bring about a greater sense of connection in general, evidenced in increased pro-social or pro-environmental behaviours. As described in Chapter 2, the restorative effects of nature can diminish negative emotion and associated psychological effects, in turn increasing our capacity to engage in positive interpersonal and social interactions. Johansson, Hartig, and Staats (2011) argue that both social contact and nature exposure restore some psychological resource, which, when satisfied by one form of connection, does not require or result in further satisfaction by the other. This suggests a ceiling effect for how 'restored' one can be. Their study used the World Health Organisation measure (WHO-5) of five statements to assess subjective wellbeing: 'Over the last seven days…: 1. I have felt cheerful and in good spirits; 2. I have felt calm and relaxed; 3. I have felt active and vigorous; 4. I woke up feeling refreshed and rested; 5. My daily life has been filled with things that interest me'; to which people could respond on a scale from 0 (None of the time) to 5 (All of the time). Social connectedness was measured by questions: 'Over the last 7 days, roughly how often have you met with a person or people who are important to you (i.e., friends, family) who do not live with you?'. Nature exposure was assessed by asking how many times they had visited it in last 7 days, and how much nearby nature was visible from participants' home or visible in the neighbourhood with options from 1 (not at all natural) to 7 (completely natural). From this combination of data, the authors (ibid.) conclude that for those living in greener areas a relative lack of social connectedness was less important for subjective wellbeing—biophilic experience somehow compensating for poor social connection.

Nevertheless, much of the wellbeing derived from woodland appears to have a social component. Access to wellbeing across the GfW case studies was frequently collaboratively constructed with other people—peer and family groups or temporary communities of interest or practice. Although sometimes short-lived or reinforcements of existing interpersonal relationships, these groups often strongly supported achievement of personal and social goals. Case studies highlighted different examples of how interpersonal connection could be both a process and outcome central to woodland activity. Continuing the theme of 'difference' or 'being away' within the research (a combination of nature's restorative effects and the different social and cultural environment), the contrasts that woodland activities and settings presented to people's everyday social experience sometimes had a significant influence on interpersonal connections. In this chapter, we take a closer look at case studies that demonstrate ways that social wellbeing was promoted by time and activities in woodland. The first two detailed case studies are from research undertaken by the Forest of Avon Trust.

The Researchers' Organisation

The charity, Forest of Avon Trust (FoAT), plants trees, provides training and advice and works with children, businesses and communities through tree planting, social forestry and Forest School delivery across the former county of Avon. FoAT working in partnership with the practitioner-researcher's Into the Woods were seeking to encourage people to visit, with support, their local wooded areas, connect with nature and build confidence to explore other woods independently. Into the Woods works especially with groups and individuals who may not normally visit woods, including young people and adults with learning or physical disabilities and mental health and dementia service users.

Case Study 1: Sharing Success

The first piece of practitioner research focused on in this chapter involved eight young adults between 17 and 24 years of age (five male, three female). All had learning difficulties and were attending a further education course designed to support school leavers from special schools or supported within mainstream education to develop independence and skills for adult life. Students from this course attended six weekly FoAT woodland activity sessions, similar to Forest School and including games, arts and crafts, fire lighting, cooking and practical conservation work.

The leaders of the sessions, Nicola Ramsden and Rachel Tomlinson, were the practitioner-researchers, who wanted to investigate the potential impacts on health and wellbeing of these sessions and understand what led to those impacts to inform best practice and session design. Evidence of the impact was also useful to persuade potential partner organisations, such as colleges and day care centres, to invest in this provision.

The Woodland Site

Into the Woods, Forest School-style sessions, were intended to build students': resilience; physical confidence and stamina; personal safety skills around fires, tools etc.; teamwork; and knowledge and appreciation of the natural world. They included games, sensory, art and creative activities, as well as practical tasks such as coppicing, fire lighting and cooking. All took place in New Barn Wood, an enclosed area of woodland surrounded by the open parkland of Ashton Court Estate, a public park owned by Bristol City Council. The woodland includes mature trees such as sycamore and oak and smaller trees and shrubs such as holly and hazel. Participants travelled to the site in a minibus and had a packed lunch in the woods each week.

Staff both at the college and within the Forest of Avon Trust expected that participation would enable the young people to develop themselves, feel safe and supported within and through social relationships and competent (and be recognised by others as competent). College staff primarily focused on individual personal and social development opportunities in keeping with the aims of the further education course students were attending.

The practitioner-researchers were sensitive to the fact that many students taking part couldn't easily articulate their ideas, thoughts and feelings. They chose data collection methods to help address this, including adult-mediated information from pre-programme interviews with college staff, a midway interview with the course tutor and staff-completed feedback sheets about what participants had enjoyed or not and what they would like to do next time. Creative audio-visual methods, such as: a pre-programme slide show and discussion to gather student expectations of the woodland experiences; card sorting where potential participants looked at photos of different woodland features or activities and chose those they liked or did not like; scrapbooks where each participant drew, wrote and stuck in things they had collected or made during sessions; feedback games like woodland charades where one person would mime one of the day's activities and others had to guess what it was; a blackboard on which anyone could write or draw to express their views and feelings; and videos taken by staff and young adults provided additional insights into how experiences had been received. Nicola and Rachel also had regular discussions after each weekly session, reviewed video footage and kept reflective diaries (Fig. 6.1).

Fig. 6.1 Student drawings on the blackboard (*Source* Into The Woods)

Combining Research and Delivery

Several methods particularly helped the practitioner-researchers, Rachel and Nicola, to conduct research simultaneously with their delivery. They scribbled down observations about and quotes from students directly onto sticker sheets labelled with participants' names, during sessions. At the end of the six weeks, the stickers for each student were brought together providing a strong picture of their journey over the full course of sessions. Tutors had to work hard to get an even spread of comments, however, and not let characters or moments dominate observations. Young people's videos allowed some of them to express themselves more fully, indeed, one participant would speak alone to the camera, but rarely in the group. Staff made observational videos, which were helpful as they enabled tutors to review what had happened while not in the thick of leading activities and supported making reflections and interpretations. In a similar fashion, both the diaries and post-session discussion between the practitioner-researchers helped tease out interesting aspects from the research and aided critical appraisal of activities and associations with outcomes. A further layer of critical reflection was provided by a midway interview between tutors and the course leader who reported on changes possibly attributed to time spent in woodland activities.

> Everyone is playing music on sticks, with leaves etc. People are singing along in a circle and seem to be enjoying themselves, there is laughter (especially from CP3) (CBC03_video_01).

> CP8 is really pleased with his quick lighting of the fire and wants others to see. He is very pleased with his success, 'Look at that one, it's still burning!' (CBC06_SV1_observations)

Sticking Together

The aspect of social wellbeing identified as an anticipated outcome prior to sessions was that students would feel safe and supported within and through social relationships. Staff members connected to the referring organisation discussed this most frequently in terms of how the group would probably 'stick together' in the woods or how they would be responsible for keeping the participants safe. Most frequently social support was considered in terms of how the woods might be made to feel secure and accessible for students likely to be relatively unfamiliar with that environment.

> They may need support with activities in terms of explanation with what the activity involves, but after a couple of weeks if the activity goes on for more than one or two weeks again [we] may be able to withdraw support…in terms of the number of prompts that may be required for a student, it could be the level of observation that a student requires (CS2 pre transcript).

Interestingly, social wellbeing was often linked by the practitioner-researchers to collective games and 'play', social and fun activities. This echoes a tension sometimes observed in other outdoor learning research where enjoyment is not always seen as foundational to learning experiences. However, there is also a body of knowledge that acknowledges enjoyment as implicated in *engagement* and hence to learning (Waite, Passy, Gilchrist, Hunt, & Blackwell, 2016). Might it be that in this context, positive emotions and moods were initially overlooked as a pathway towards engagement, overshadowed by concerns about an unfamiliar environment.

> CP6 is quite reluctant to engage normally [at college] so the fact that he is engaged when he's in there [the woods] is good (CBCmid_interview_CS1_F1_F2_01).

Rachel and Nicola evidence positive emotional outcomes to be part of a virtuous circle, both leading to and resulting from moments of group cohesion. Feeling good helps the group to engage in collective interaction, activity and support and this in turn helps the group to feel good. This ties in with Fredrickson's 'broaden and build' model of positive emotion as facilitating experimentation with new relationships and behaviours, explored in Chapter 2. Their observations affirm that positive emotion can be closely affiliated with other forms of wellbeing in these feedback loops. The practitioner-researchers felt that the collective experiences they observed, combining social and emotional wellbeing, could not be mapped onto GfW's existing framework and constructed an indicator called 'connecting with others through shared experience'. They found this to be the second most common experience of woodland wellbeing after individual experiences of positive emotion. It is similar to the indicator of psychological wellbeing 'connecting with others through shared beliefs and outlook' (explored in case study 1, Chapter 7) but emphasises shared activity.

> Bonding through a shared experience leads to well-being within a group…It is the sharing of an experience that make people feel part of a group - they have a common background. It's about feeling a group identity, and relating to people through and because of this, because of the things that you've done together… 'shared experience' is more about social and emotional wellbeing. Connecting through shared experience comes from the situation rather than the people (Rachel and Nicola PR's).

Over half of the instances of this indicator were about 'what' people were doing. A shared experience might also enable empathy with others.

> Several of the male students started singing a football-style chant (to do with the food) and seemed to be really enjoying this as a boisterous group activity (CBC06_F2_diary).

> Playing ring on a string - in between laughing, it goes quiet as people concentrate but they are smiling with expectation (CBC04_play_06 video).

> CP1 asks 'what next' and talks about games they have played at forest school she has enjoyed. The group move together to discuss what game should follow hide and seek (CBC06_SV1_observations).

Social wellbeing was sometimes located within students' processes of familiarising themselves with new places or routines, becoming a more independent and active group. Sometimes shared recollection (as above) reinforced these positive feelings.

> I really noticed on the way to the wood from the bus, he (CP6) wasn't the last [anymore] … he was up with the rest of them and he wasn't with a member of staff, and he didn't need anyone (CBC05_F1_F2_discussion01).

> CP2 was reminiscing about last week when she and CP6 had played a game where they drew a line in the ground. They also chatted about which way was best into the woods (CBC06_F1_diary).

Other more common outcomes of time spent in the woods included social dimensions of other forms of wellbeing. Students demonstrated feelings of competency (psychological wellbeing) and also enjoyed being seen by others to be competent during the six weeks.

> CP5 was really pleased with lighting the fire actually. When she came up to wash her hands she went "I was brilliant!" (CBC05_F1_F2_discussion01)

> CP8 is really pleased with his quick lighting of the fire and wants others to see. He is very pleased with his success, 'Look at that one, it's still burning!' (CBC06_SV1_observations)

> CP4 wants everyone to see that he has lit birch bark successfully, after lighting the cotton wool very quickly. He lights it again and again (CBC06_SV1_observations).

These were often linked to practical tasks including cooking, tending the fire, using tools and whittling sticks. It seemed that this competence developed through repetition of activities and gradually becoming more confident at taking control over them. Following Forest School practice of small steps towards achievement, activities were deliberately chosen/set-up so that initial challenge gradually led to a sense of achievement. The young adults also clearly welcomed the opportunity for free play which opened the way for them to connect with others socially and feel and be seen to be competent (through bringing their own ideas to sessions, organising games and sharing these with peers for example). A balance between structured and unstructured activity was clearly important in maintaining learner-centred practice so that young people could achieve their own wellbeing needs (Waite & Davis, 2007).

Implications for Practice

When following up the results of the programme a year later, college staff said that the young people had interacted with each other in the woods in a way that they didn't in college. For example, students would generally not sit together for lunch in college but were happy to do so in the woods. The lack of transfer of social wellbeing to the college context is disappointing if it means that the social wellbeing is only being experienced in the bubble of woodland activity, but a problem of transfer of beneficial outcomes is consistent with other forms of outdoor learning (Brown, 2010) and with our own observations about young people in the Embercombe YLP, described in Chapter 5, once they returned to their home circumstances. Possibly this is because the intangible ingredient of the engaging and restoring natural world that supports these experiences is absent from their usual social settings. However, general dispositions towards positive social wellbeing in other contexts may be enhanced through long-term, rather than short or one-off, woodland sessions, where people are then able to build shared experiences and memories that can act as a relational resource in different environments.

Rachel and Nicola found it tricky to work out specific roots of wellbeing. Stakeholders tended to talk in the abstract and focus on the environmental aspects of the 'place' (the woods) rather than specific activities; while their research analysis focused on the activities themselves, such as collecting wood, and individual components of 'being in the woods' such as uneven ground, weather, mud, individual relationships in the group. The difference in holistic and nuanced approaches may be a result of different levels of engagement with the natural environment between college and woodland activity staff but taken together, they provided insights which one or other approach would not have yielded.

Finding out what initial expectations are, even literally where the group is coming from, can help frame woodland activities in a more nuanced way. Both transfer and deeper insights from alternative perspectives may be enhanced by close collaboration between the usual setting and the woodland. Accompanying the group from the setting forms a bridge between the contexts that may make sharing feelings and transfer of a sense of wellbeing more likely. The use of scrapbooks can also act as a tangible and valued vehicle for better communication and aid sharing ideas across both contexts. These scrapbooks were personal for each young person, but they could be made dialogic so that feedback is two or three-way between the woodland activity leaders, the participants and the usual setting. In this way, they could double as a platform for recognition of achievement by others, a factor that seemed to be so important to this group of young adults with learning difficulties.

Top 3 messages from 'Into the Woods' for young adults with learning difficulties

1. The young people experienced positive emotions and moods (emotional wellbeing) through being in a group (social wellbeing). The opportunity to play and devise games, having autonomy to act, was seen as a key contribution. Creating opportunities for self-determination is essential.
2. Developing themselves, succeeding in new tasks, becoming competent and showing others were found to be important parts of the young adults' wellbeing. The process of small steps to achievement is fundamental to Forest School practice. However, with this group, the recognition of achievement by others seemed an important factor.
3. Connecting with and being with others as a group was considered significant (in follow-up research), particularly by the college staff. This highlights the importance of good liaison with other organisations to determine priorities to work on, aspects of the process that are influential, and key outcomes.

Case Study 2: Tackling Tasks Together

The second piece of research carried out by Rachel and Nicola was an 'Into the Woods' project with an older group of ten adults with learning and physical differences (two were wheelchair users), referred from a day centre in Bristol (8 male, 2 female). Again the practitioner-researchers wanted to find impacts on health and wellbeing from woodland activities provided in Forest School-style sessions and to understand what led to these impacts. Two women and eight men aged between

21 and 64 (with 3–4 support staff) attended the programme for one day per week over six weeks, travelling by minibus from their day care centre.

The Woodland Site

Sessions again took place at several wooded locations on the Ashton Court Estate, parkland open to the general public. Areas were composed of mixed woodland of various ages within the estate with easy level walking access from the parking area for those using wheelchairs or walking aids. The group brought a packed lunch to eat in the woods, hot drinks were provided, and a fire was made on several occasions for warmth and cooking. Other activities included feedback games, listening exercises, collecting colours from nature, blindfolded 'meet a tree', mud painting, clay modelling and making music. The group also carried out practical tasks such as coppicing and making a windbreak.

Staff at the day centre generally shared a belief that being in woods was a good thing for health and wellbeing through enabling close contact with the natural world. Many also suggested that wellbeing would be enhanced through physical or personal development associated with the way a new situation would challenge or stretch participants (including the unfamiliar environment, trying new activities or being with people they didn't know).

> There's something you get from being in woodland, or things that I do, it's more immersive… when you're in a forest you're more closed in so you're more inclined to look at the finer detail (shire_preinterview_SS4_transcript).

> I think people can get bored here [day centre] and it's a different activity—so it's an activity that is unusual (Shire_preinterview_SS1SS2_transcript).

> People might help each other more in the way they wouldn't in a more familiar environment (shirepre_interview_group_SS3_SS4_F1diary).

Connected to this was an expectation that the group would learn skills and feel competent. Staff also anticipated that the group would benefit from being with other people in the wood, because they would be working and learning together as a shared experience, and importantly, would feel supported by the rest of the group in new and unfamiliar situations. Finally, staff expected those involved in the programme to feel relaxed and peaceful, which was almost always attributed to the woodland itself. In this group, it was the woodland or natural environment that was seen to be of most importance and stakeholders resisted any finer focus on specific activities.

As with the younger age group, the practitioner-researchers used photo card sorting and slideshow prompted discussion with students, alongside interviews with staff, to find out expectations before the programme. Similarly, they repeated the use of videos, sticker sheets, blackboard, scrapbooks and also employed feedback games to collect data. Day centre staff completed feedback questions about what participants had enjoyed or not and what they would like to do next time. Again, post-session discussion between practitioner-researchers and personal reflective journals added further layers of interpretation.

Safe Supportive Relationships in Shared Community

Safe and supported within and through social relationships was the most frequently recorded indication of woodland wellbeing within this case study. It was frequently recorded when staff and carers were working alongside providing positive feedback to participants during an activity or emerging routines, such as making their way to and from the woodland site. It was also recorded sometimes when the whole group was involved in an activity, such as sitting around the fire eating lunch. Many of the adults needed practical help with tasks, but also highly appreciated support in the form of reassurance and reflection. This came from care staff, but also in the form of mutual support amongst the group as participants checked on or encouraged each other from time to time (Fig. 6.2).

> SS3 leads SP1, blindfolded, to a tree, gets him to feel the trunk, asks him if it's big, leads him away again, turns him around and takes off blindfold. SP1 heads straight to his tree and touches it. SS1 "How d'you know that,

Fig. 6.2 Students gathered together around the fire during Into The Woods activities (*Source* Into The Woods)

> SP1?" SP1 smooths trunk and says "soft". F1 "Brilliant" SS3 "Well done mate" (video shire05_meetatree_03).
>
> As the activities draw to a close the students check in with each other, using thumbs up signals to ask if the others are fine and calling across the fire circle to ask if each other are okay (Shire06_observation_SV1).
>
> SS3 is helping SP4 to make his leaf crown, they both seem quite absorbed in this. SP3 is laughing in the background and waves at SS4 (video shire06_leafcrowns_05).

Emotional wellbeing was the second most frequently observed indication of woodland wellbeing, often associated with creative and sensory activity. It was next most commonly associated with interactions with leaders and the group.

(SP7 is wearing his crown and smiling) SV1: 'how do you feel when you wear that?' SP7: 'happy', SV1: 'what do you think when you see everyone else wearing theirs?' SP7: 'happy' (video shire06_leafcrowns_06).

Being seen to be competent, the social dimension of feeling competent (psychological wellbeing) again emerged as an important dimension of woodland wellbeing for this group. For example, people participating in practical and creative activities were able to demonstrate their skills and abilities to others. It was also experienced when service users could see that they had carried out a task according to instructions and were meeting the expectations of the leader. Seeking acknowledgement of achievements could be easily recorded.

I'm going to tell people what I did today—cutting trees down (Shirelink_SP2_stickers).

Made a boat-shaped Blobster "Look—I've done it!" (Shirelink_SP8_stickers)

At other times, Rachel and Nicola noted that creative activities led to a sense of flow—absorption in and rising to the challenge of the task—and experiences of competency that were independent of being witnessed. However frequent outcomes such as having a sense of purpose and feeling competent (most often associated in the study with learning new skills and working towards and achieving a goal) were enabled by social support from tutors and feelings of social safety. Unfamiliar woodland surroundings and novel, engaging activities in fact intensified interpersonal relation between staff and service users. This was not always because group leaders and centre staff were always on hand to help and encourage participants to have a go at new things (they also stood back and encouraged participants to try things independently). It also originated in the fact that centre staff were encouraged to experiment with and take part in woodland activities on an equal footing with adult service users. In follow-up interviews with Shirelink staff six months and one year later, it emerged that this co-participation had been highly valued by staff as an opportunity to learn new things about the people they cared for and view them in a different light.

In terms of aspects of social wellbeing identified by Teghe and Rendell (2005), this helped staff to accept others as they are and feel part of a shared community. For the adults with learning difficulties, they experienced different relationships within the new woodland community of practice enabled by cultural expectations there (Adams, 2012; Waite, 2013). There was a shift within usual patterns of helper and helped, and through feeling, they were contributing to the community and culture in the woodland site adult service users could experience social integration, social contribution and social actualisation.

The practitioner-researchers identified that social relationships underpinned most activity in the programme acting as a foundation for other forms of wellbeing and/or associating with it in the kind of positive feedback loops identified within their social-emotional indicator 'Connecting with others through shared experience'.

Top 3 messages from 'Into the Woods' for adults aged between 21 and 65 with learning difficulties

1. Social relationships in the group were key to the wellbeing of the participants, especially in unfamiliar situations. Leaders and staff helped participants but also stood back, encouraging participants to try things independently. Staff were encouraged to take part in activities on an equal footing as participants, which allowed them to see participants in a new light. Participants were able to demonstrate their achievements to staff and peers, receive positive feedback and give support to others in the group.
2. Sensory and creative activities were important as they both appeared to allow participants to access a fundamental source of wellbeing, which may relate to flow and/or mindfulness.
3. Having a sense of purpose and feeling of being competent were key outcomes of the sessions. These are most often associated with creative activities and practical tasks, which encouraged participants to learn new skills and to work towards and achieve a goal.

Nicola and Rachel pointed out strong parallels between their use and adaption of GfW indicators and the new economics foundation's '5 Ways to Mental Wellbeing' (Nef, 2008), which have been promoted widely by

Public Health England, the NHS and other health and wellbeing agencies. They propose that the 5 actions can act as a framework for planning woodland activities that can promote wellbeing in groups of staff and service users with learning/physical differences.

Points for practice

CONNECT: Feel safe & supported
Ensure there are enough staff to offer adequate support. Support staff and carers should be briefed and participate fully in all activities as an equal member of the group and encourage service users to try things independently and together, only giving help when needed.

BE ACTIVE: Sense/mindfulness/practical physical tasks
Incorporate lots of sensory and hands on activities to enable experience of activities and sensations that might rarely be available to them.

TAKE NOTICE: Engagement & absorption
Ensure that plenty of creative activities are included in the programme and that time and space is given to allow everyone to explore their own creative urges.

GIVE: Purpose and responsibility
Give people tasks with clear outcomes or goals to work towards and entrust them with responsibility for their completion. Repeating a routine activity, such as setting up the fire circle and involving people in helping can give participants the confidence to initiate the activity themselves over time.

KEEP LEARNING: Feel competent
Adapt activities so that all participants can achieve (for example, provide suitable tools, make workspaces suitable for wheelchair users). Encourage peer learning and allow participants to take control of sessions. Make opportunities for positive feedback, recognition and acknowledgement. Include quieter participants, who are less likely to come forward, in these opportunities.
Adapted by the practitioner-researchers from 5 ways to mental wellbeing, Nef (2008).

The Forest of Avon Trust has continued to be involved in researching the benefits in similar projects with targeted groups. For example, evidence from their 'Woodland Wellbeing' project for people with dementia (Gibson, Ramsden, Tomlinson, & Jones, 2017) echoes the impact on social wellbeing established in their earlier GfW case study, finding lower levels of isolation and loneliness amongst participants following a woodland activity programme. Their improved understanding arising through their careful evidencing continues to nuance their recommendations for practice with different groups.

References

Adams, M. (2012). A social engagement: How ecopsychology can benefit from dialogue with the social sciences. *Ecopsychology, 4*(3), 216–222.

Burgess, J. (2002). 'But is it worth taking the risk?' How women negotiate access to urban woodland—A case study. In R. Ainley (Ed.), *New frontiers of space, bodies and gender* (pp. 133–146). London: Routledge.

Brown, M. (2010). Transfer: Outdoor adventure education's Achilles heel? Changing participation as a viable option. *Australian Journal of Outdoor Education., 14*(1), 13–22.

Cameron-Faulkner, T., Melville, J., & Gattis, M. (2018). Responding to nature: Natural environments improve parent-child communication. *Journal of Environmental Psychology, 59*, 9–15.

Cartwright, B. D. S., White, M. P., & Clitherow, T. J. (2018). Nearby nature 'buffers' the effect of low social connectedness on adult subjective wellbeing over the last 7 days. *International Journal of Environmental Research in Public Health, 15*, 1238.

Dadvand, P., Hariri, S., Abbasi, B., Heshmat, R., Qorbani, M., Motlagh, M. E., … Kelishadi, R. (2019). Use of green spaces, self-satisfaction and social contacts in adolescents: A population-based CASPIAN-V study. *Environmental Research, 168*, 171–177.

Forest Europe. (2019). *Human health and sustainable forest management.* Bratislava: Forest Europe Liaison Unit.

Frazer, D. (2017, September 22–24). *Can you be happy when others aren't?* Workshop at Building Wellbeing Together, Stroud.

Goodenough, A. (2015). *Social cohesion and wellbeing deriving from woodland activities: Good from woods* (A Research Report to the BIG Lottery).

Goodenough, A., Waite, S., & Bartlett, J. (2015). Families in the forest: Guilt trips, bonding moments and potential springboards. *Annals of Leisure Research, 18*(3), 377–396.
Gibson, E., Ramsden, N., Tomlinson, R., & Jones, C. (2017). Woodland Well-being: a pilot for people with dementia. *Working with Older People, 21* (3), 178–185. Retrieved from https://doi.org/10.1108/WWOP-05-2017-0012.
Helliwell, J. F., & Putnam, R. D. (2004). The social context of well-being. *Philosophical Transactions of the Royal Society London B., 359,* 1435–1446.
Johansson, M., Hartig, T., & Staats, H. (2011). Psychological benefits of walking: Moderation by company and outdoor environment. *Applied Psychological Health and Well-Being, 3,* 261–280.
Kasper, D. V. S. (2009). Ecological habitus: Toward a better understanding of socioecological relations. *Organization & Environment, 22,* 311–326.
Maller, C., Townsend, M., Pryor, A., Brown, P., & St Leger, L. (2006). Healthy nature healthy people: 'Contact with nature' as an upstream health promotion intervention for populations. *Health Promotion International, 21*(1), 45–54.
Morris, J., O'Brien, L., Ambrose-Oji, B., Lawrence, A., & Carter, C. (2011). Access for all? Barriers to accessing woodlands and forests in Britain. *Local Environment, 16* (4), 375–396.
New Economics Foundation. (2008). *Five ways to wellbeing.* Retrieved from https://neweconomics.org/2008/10/five-ways-to-wellbeing-the-evidence.
New Economics Foundation. (2011). *Five ways to wellbeing: New applications, new ways of thinking.* Retrieved from https://neweconomics.org/uploads/files/d80eba95560c09605d_uzm6b1n6a.pdf.
NICE. (2019). *Social and emotional wellbeing in secondary education.* Retrieved from https://pathways.nice.org.uk/pathways/social-and-emotional-wellbeing-for-children-and-young-people/social-and-emotional-wellbeing-in-secondary-education.pdf.
O'Brien, L., Townsend, M., & Ebden, M. (2008). *'I like to think when I'm gone, I will have left this a better place': Environmental volunteering: motivations, barriers and benefits.* Farnham: Forest Research.
O'Brien, L., & Morris, J. (2014). Well-being for all? The social distribution of benefits gained from woodlands and forests in Britain. *Local Environment, 19*(4), 356–383.
Rogerson, M., Barton, J., Bragg, R., & Pretty, J. (2017). *The health and wellbeing impacts of volunteering with the wildlife trusts.* Newark: The Wildlife Trusts. Retrieved from https://www.wildlifetrusts.org/sites/default/files/2018-05/r3_the_health_and_wellbeing_impacts_of_volunteering_with_the_wildlife_trusts_-_university_of_essex_report_3_0.pdf

Stigsdotter, U. K., Palsdottir, A. M., Burls, A., Chermaz, A., Ferrini, F., & Grahn, P. (2011). Nature-based Therapeutic Interventions (Chapter 11). In K. Nilssen et al. (Eds.), *Forests, trees and human health*. New York: Springer.
Vering, K. (2006). Social sustainability—Forest projects for the integration of marginal groups. *Urban Forestry & Urban Greening, 5*(1), 45–51.
Waite, S. (2013). 'Knowing your place in the world': How place and culture support and obstruct educational aims. *Cambridge Journal of Education, 43*(4), 413–434.
Waite, S., & Davis, B. (2007). The contribution of free play and structured activities in Forest School to learning beyond cognition: An English case. In B. Ravn & N. Kryger (Eds.), *Learning beyond Cognition* (pp. 257–274), Copenhagen: The Danish University of Education.
Waite, S., Passy, R., Gilchrist, M., Hunt, A., & Blackwell, I. (2016). *Natural Connections Demonstration Project 2012–2016* (Final report). Natural England Commissioned report NECR215. Retrieved from http://publications.naturalengland.org.uk/publication/6636651036540928.
What Works Wellbeing. (2018). *Places, spaces, people and wellbeing: Full review*. Available online at https://whatworkswellbeing.org/product/places-spaces-people-and-wellbeing/.

7

Natural Sources of Psychological Wellbeing

What Is Psychological Wellbeing?

The adjective 'psychological' suggests that this type of wellbeing is principally located within the mind and in contrast to emotional wellbeing; the emphasis is more on positive cognitive faculties and a sense of competence in meeting the demands of tasks. It is aligned to eudemonic rather than hedonic wellbeing and refers to a satisfaction with one's ability to function well in the world. The Good from Woods (GfW) working definition associated psychological wellbeing with feelings of being in control, competent (and seen by others to be competent), energetic, purposeful, developing oneself, connecting with others through shared beliefs and outlook and secure with personal limitations.

A. Goodenough and S. Waite, *Wellbeing from Woodland*,
https://doi.org/10.1007/978-3-030-32629-6_7

Feeling on Top of Things

One common symptom of poor mental health is a sense that everything is getting on top of you and that your ability to cope with what life throws at you has been compromised (Sonntag-Östrom et al., 2015; Stigsdotter & Grahn, 2011). This lack of confidence in managing day-to-day life is associated with both depression and anxiety. The dimension of psychological wellbeing most frequently under exploration in studies of the effects of natural environments appears to be emotional regulation and our management and experiences of negative emotions such as stress, depression and anxiety. However, as explored in Chapter 3, psychological wellbeing in our understanding of woodland wellbeing is also associated with our ability to actively function positively.

Green space has been recognised as contributory to better mental health (Seymour, 2003) but studies on the psychological aspects of interaction with natural spaces tend to stay firmly within disciplinary boundaries. However, as we have noted elsewhere, the complexity of human/nature relationships increasingly demand more integrated cross-disciplinary approaches. Sanesi, Lafortezza, Bonnes, and Carrus (2006), for example, with backgrounds in environmental psychology and urban forestry, compared studies of the social and psychological dimensions of urban green space use conducted from these different disciplinary perspectives. Both used survey methods, but urban forestry was more descriptive and underpinned by an ecosystem services viewpoint, such as mitigation of pollution. The environmental psychology case sought to understand aspects of human–environment relationships that might account for the attitudes expressed through sophisticated statistical analysis. The authors found that although findings were broadly comparable and mutually supportive, deeper insight and guidance for green space design would be possible through integrated multidisciplinary research. The Sanesi et al. (ibid) comparision research found green appreciation was more prevalent amongst those with the most or least access, which the authors attribute to greater familiarity on the one hand and greater need on the other. The authors conclude that understanding the impact of green space on psychological wellbeing requires further interdisciplinary collaboration.

Restorative Settings

Research has explored how restorative effects of natural settings (as detailed in Chapter 2), and nature relatedness may support psychological wellbeing. Sobko, Jia, and Brown (2018) focusing on relatedness combined results of two scales (an adapted form of the Connectedness to Nature Index-CNI and the Strengths and Difficulties Questionnaire-SDQ), completed by parents of preschool children in Hong Kong to evaluate relationships between children's nature connectedness and their psychological wellbeing. They found that overall the greater the child's perceived connection to nature, the healthier the child was reported to be. Feeling a 'responsibility towards nature' was most predictive of positive psychological wellbeing, which links well with our exploration of young tree planters' feelings of wellbeing discussed in Chapter 9.

The converse of this relationship might be seen in studies showing that how well we are coping psychologically appears to affect our view of nature and what we need from it. Peschardt and Stigsdotter (2013) find interesting distinctions in what was perceived as important for restoration in small public urban green spaces in Copenhagen. The authors (ibid.: 27) establish that for 'average' users of these space the sensory dimensions of 'serene (silent and calm)' and 'social (entertainment and restaurants)' green areas were perceived as most restorative. More stressed individuals however also rated 'nature (wild and untouched)' space as having relatively high restorative properties. Overall Stigsdotter, Corazon, Sidenius, Refshauge, and Grahn (2017: 2 citing, Grahn and Stigsdotter, 2010) find that people requiring 'psychological restoration' have a preference for natural settings that are; serene 'a haven, almost a holy place'; a refuge 'where people can feel safe'; rich in species 'diverse in sensory experiences'; and nature 'wild, free-growing, untouched room'.

Stigsdotter and Grahn (2011: 299–300) also found that more stressed individuals preferred certain types of activity in outdoor space, as compared with those less stressed. Activities which don't involve other humans but do often require engagement with the natural world and its inhabitants were preferred by the more stressed: rest activities (including 'getting fresh air', 'watching wild plants'); followed by animal activities (interspecies interactions such as 'studying pets' and 'feeding animals'); and walking activities (including 'enjoying the verdant' and 'getting fresh air'). Less stressed people preferred more social activities, while younger individuals seemed to prefer more physical activity.

A Swedish study of a three-month forest rehabilitation programme for patients with exhaustion disorder (Sonntag-Östrom et al., 2015) explored the process through interviews. Patients reported 'striving for serenity' was challenging in their first experiences of two hours of solitude and inactivity in the novel surroundings of the woods (ibid.: 609). As they gradually became familiar with the new context, they found 'peace of mind' in favourite places within the forest environments. Favourite places were characterised by openness, light and a good view, while also shielded from others' view, and were sites where patients began to rest and reflect. This again suggests psychological wellbeing achieved through an intensification of interactions with natural space. Over time, most reported positive changes in their home life and began to make plans for the future and test out new behaviours. The social gatherings at the beginning and end of each forest visit were stressful for some, and many reported growing feelings of anxiety as the programme ended and return to their everyday life was imminent. The researchers suggested that the forest alone was insufficient to effect change and that cognitive behavioural therapy alongside these experiences would speed recovery, but also propose woodland experience is sustained for participants as they progress on their road to psychological wellbeing. Von Lindern (2015) notes that where there are more correspondences between daily life and the nature visit, the less the experience is seen as restorative. This supports observations made in Chapter 2 regarding the comparative cultural 'lightness' of many outdoor settings (Waite, 2013) and observations across case studies of the psychological value of 'being away' from everyday demands. It also emphasises that the domain of social wellbeing and social aspects may be less beneficial for those who are struggling to cope with their lives at least initially.

Place and People Interactions

Variations in the effects of nature on health and happiness are also established by Herzog and Strevey (2008), who distinguish between factors associated with psychological and emotional wellbeing. The authors find in their study that while a sense of humour predicts emotional wellbeing and reduction in feelings of stress, contact with nature is more closely associated with psychological wellbeing or functioning well. This would suggest that the social aspects of woodland activities, which often include fun enjoyable elements, link to emotionally feeling good, but feeling competent and purposeful is mediated by aspects of the material natural context. Attention restoration theory would propose that this is due to a direct impact of soft fascination relieving cognitive load.

Stigsdotter et al. (2017) report on a study on Health Forest Octovia, where different forest environment qualities have been deliberately manipulated to represent eight perceived sensory dimensions (some of which are explored in related research above): social, prospect, rich in species, serene, culture, space, nature and refuge. They found individuals' past experience of natural environments was influential in the favouring of serene, followed by rich in species, refuge and nature dimensions for restoration (ibid.: 2). Tyrvaïnen et al. (2013) reported gender differences with restorative effects of nature most pronounced for healthy middle-aged women, while Zhang, Howell, and Iyer (2014) found that individuals who were emotionally attuned to nature gained greater benefits in psychological wellbeing through nature connectedness. This points to the interrelated aspects and entanglement between our biography, biology, culture and capacity to achieve wellbeing benefits from nature and the woods. It may indicate that a stepping stones approach is likely more effective for those who are more stressed and/or have little previous experience of and attachment to nature.

Factors Associated with Positive Functioning

Across the various research projects in GfW, psychological wellbeing was reported widely and associated with our pilot indicators of feeling: in control, competent (and seen by others to be competent), energetic, purposeful, developing oneself, connecting with others through shared beliefs and outlook, secure with personal limitations. This type of wellbeing aligns most closely with 'eudemonic' wellbeing as described in Chapter 3, as it is about the functioning of the individual and the agency and control, they experience in shaping environments, material and social, for their purposes. This may be because many of the activities studied focused on outdoor learning and opportunities for skills development intended to have psychological impacts on functioning in certain contexts. In addition, psychological wellbeing as an immediate outcome is perhaps easier for participants to articulate or for others to notice, like emotional wellbeing, rather than more nuanced impacts. In the following two case studies, we examine some of the specific associations associated with psychological wellbeing.

Case Study 1: Taking Action, Changing Cultures

This case study explored the activity of coppicing at Ruskin Mill College with 13 students (12 male: 1 female) that were taking part in this activity as part of their school curriculum for up to three days a week. Student experience in the coppice ranged from play (making dens, exploring, making dams) to felling trees with hand tools and studying for chainsaw qualifications. Coppicing is an important part of the yearly cycle of woodland management at the college and valued for education and therapy. The college-owned woods are part of a wider learning community and a range of interested people including teachers, learners, land workers, craftsmen, therapists and artists were involved as tutors or helpers in sessions. The main practitioner-researcher in this case study was a Ruskin Mill College tutor involved in the coppice activity, Richard Turley, supported by social forester Rachel Tomlinson (who had previously researched the Into the Woods projects, Chapter 6).

The Researcher's Organisation

Ruskin Mill College (RMC) is located in a Gloucestershire valley and is centred around two restored eighteenth-century woollen mills containing craft workshops and administrative centres. The valley bottom contains a market garden and trout fishery, and the top of the valley accommodates a 100-acre farm with livestock and horticulture. It is part of a larger national organisation of 5 similar colleges, Ruskin Mill Trust (RMT) that works with students with learning difficulties aged 16–19 years. RMT colleges aim to advance the education of young people with learning, social and/or behavioural needs through training in the areas of arts, crafts, agriculture and environmental sciences, influenced by the ideas of Rudolph Steiner, William Morris and John Ruskin about the value of practical work and natural surroundings. The college has been offering land-based experiential education for more than thirty years through activities identified as meaningful, purposeful and grounded in the geographical and historical context of their environment.

The Woodland Site

The steep valley sides include twenty-two acres of ancient beech woodland in which, at the time of the research, there were 10 areas of coppice, one felled every year. The case study site had been coppiced 10 years previously and planted with new hazel. The hazel stools were small and undeveloped. Some beech and ash standards remained, and these have seeded many small whippy saplings at the east end of the coppice. The coppice is isolated with no nearby main roads and few houses. There is a natural spring nearby from which drinking water is collected. Roe deer and muntjac graze the wood and many birds are also present including the elusive woodcock as well as woodpeckers, nuthatches, treecreepers, sparrow hawks and buzzards. The coppice is a clearly defined area within the woodland. It is a space which is created and differs from other parts of the college in that it is temporary and to some extent a hidden secret, intimately known to only those who work there. Coppicing involves cutting the underwood such as hazel, ash and sycamore down to ground level and removing enough of the taller, canopy, trees to allow light in. Coppicing allows the flora to flourish and benefits the ecology and biodiversity of the area.

Tutors leading the coppicing explained that they felt this space contrasted with other parts of the college in that it was actively being created by the students and was intimately known only by those working there. In this sense, it was a special place for the group of students that were regularly helping there. It is interesting to wonder whether its qualities met some of the preferences for nature that stressed individuals prefer for psychological restoration in the studies above—serene, refuge, rich in species and nature (Stigsdotter et al., 2017). In some ways, the coppice started out as 'nature' ('wild, free-growing and untouched') (ibid.: 2) but was with the help of students as described below transformed into something closer to refuge ('where people can feel safe') (Fig. 7.1).

According to the tutors, the coppicing was not only an educational activity, but served a practical purpose of woodland management and contributed to the economy of the college, providing fuel and resources for green woodwork and blacksmithing.

The practitioner-researchers were conscious that some students rarely articulated their thoughts, feelings and responses in words, so they chose appropriate research methods, employing observation and participatory methods (drawing, making, film) alongside in-activity informal conversations with students and interviews with tutors and carers familiar with the young people.

In this chapter, we focus particularly on the psychological wellbeing indicators that were noted through this data collection.

Meaningful Activity

Tutors acknowledged that the material space of the coppice could be cold, uncomfortable and the uneven terrain was difficult for some students to move around, and they recognised that newly arrived participants struggled with these potential hurdles most of all. This is like the findings of Sonntag-Östrom et al. (2015) in their forest rehabilitation programme in that basic physical needs for warmth and shelter had to be satisfied before participants could begin to gain more psychological wellbeing.

Fig. 7.1 Picture of coppice Tea Tepee drawn during data collection (*Source* Ruskin Mill Local Partner's research evidence)

Tutors also understood the unique social space and culture of the coppice, different from that of the classroom, might challenge some students. As Waite (2015) suggests, there are advantages and disadvantages associated with congruence of natural, outdoor spaces with everyday culture. For some, it is important to provide a sense of familiarity to reduce anxiety, while for others, the lack of established routines compared to their usual cultural context clears the way to establish new patterns of behaviour. Quay (2017) explains this difference as cultureplace, a unified perception of the functions and meanings of a place from an individual perspective such as playcoppice for some, workcoppice or activecoppice for others. Further to this, while tutors believed that there was almost a certainly a job within the coppice that each student could enjoy, that they could 'find something they love' (RMS03), they were aware that other tasks may be less accessible or enjoyable for that participant. Activities and contexts were differentially responded to at an individual level (Stigsdotter et al., 2017). In eliciting stakeholder expectations, the practitioner-researcher found that although potential barriers to increased wellbeing were identified, tutors believed that difficulties also offered opportunities. Each challenge might offer the potential to master a skill, behaviour, attitude or sense of purpose that might make it manageable and contribute to a sense of being in control.

> I can remember just seeing RMP22… He didn't really want to work, had to be in that little hard space and that's really what he needed. It was too much to be out exposed in the kind of weather and the coldness, but he was there…he just slowly came out of that space, and he wanted little jobs that he could manage to do for himself you know. Like he became the faggot maker and didn't last all the time, but that was a job that was contained, and he knew how to do it, and he could work away without anybody having to interrupt or tell him or for him to feel like he wasn't capable of doing something of worth [RMSO1]

The link to challenge and purpose may explain why it was psychological wellbeing or functioning well that was more commonly identified than hedonic forms of wellbeing. The expectation of contribution, despite challenges, supported a growth mindset (Dweck, 2008) that attributes success to the effort put in. However, the coppice wasn't necessarily understood as a more challenging learning environment than the classroom. Indeed, tutors felt it offered less effortful enjoyment, which students confirmed, citing the material and social setting as helping them feel good (touching, smelling and being in trees, enjoying fresh air and the sound and heat of the fire, being with peers and tutors in the Tea Tepee, for example).

Developing the theme of cultural density (Waite, 2013) where establishing norms create a form of place-based habitus over time, practitioner-researchers noted that coppicing had its own pattern that could perhaps contribute to psychological benefits for new coppicers. It might confer a sense of purpose and control dictated by nature and need. Clear practical rhythms and goals guided not only how the work was done (cutting in a way to encourage regrowth), but where (in an area of coppice regrowth), when (in the season that would allow regrowth) and why (because regrowth will yield sustainable fuel). Being part of that coppice culture also carried precise and careful expectations of how behaviour, tools and woodland need to be competently managed in harmony for safe and successful outcomes. This sense of being part and contributory to something bigger than oneself was also thought to be significant by tutors as important learning dispositions transferable to other contexts.

> …it's a job and there's this kind of sense that it has to be done for a bigger, there's a bigger need… there's this other thing that is requiring you to go through all those stages of the process even if you don't want to, that kind of sense of maybe gaining some sense of discipline from that [RMS02].

Tutors also considered the repetition of practices associated with coppicing offered more freedom for students to develop skills in and achieve activity helping them to feel purposeful, competent and capable.

> So with one of the students, for example,…for the first week he struggled just with one idea to just do routine things, but finally we got him to do two other things, I believe in the future he will be able to move to learn another skill, using another tool [RMSO3].

> One of the good things about coppice is there's always stuff to do, just until it's finished, so it's this endless work. So, all you can do is get your head down and keep working, so they are getting used to: "What we're doing today?" "We're going to the coppice", "How long are we going to be there?" "All day" [RMSO4].

Mastery

Students, the majority male, taking part in coppicing, primarily emphasised the practical skills they were developing, as contributory to them feeling good, competent and purposeful. Several students contrasted their own development and the comparative inexperience of others and seemed to derive pleasure from mastering controlled interactions with material nature via tools.

> I done splitting [today] so this is my first time really. [Long pause, and sound of wood splitting]. Yes…I like that axe, that axe is my new number one friend [Student no ID2].

> I enjoy processing wood using the billhook. I didn't think I would like coppicing, but I really enjoy it. It's better than being in a classroom [RMP01].

> Generally when you cut a tree down, you get the pleasure of when the tree falls and the job of when the first bit you do is you set everything up around and you get everything ready and then when you got the tree down, it's like good feeling you know it's down on the ground we done a good job.. (Student no ID, Rmaudio_studentsgroupdiscussion_part01)

> I'm, well I feel a little bit stronger than my first year. I am much fitter, much healthier, much stronger. And it keeps me away from my Xbox36…Yep, I know much [about coppicing], like all the first years when they come…and [in comparison] I just like I know what to do here and there and I didn't' [RMP11].

> It's all right to go fast with the axe if you know what you are doing, and you got experience [RMP10].

Observations and student films corroborated participants choosing to practice and master skills, pursue demanding processes to their completion and deriving satisfaction from their capacities and competence in the coppice.

Recognition

It's possible that the practitioner-researchers' dual role of lead researcher and tutor influenced the drift of discussions with students. Students' references to skills gain and development may partly demonstrate their awareness that these are the learning objectives and meet their tutor's expectation, but reflecting back tutor's views about the work of coppicing (correct use of tools, sorting and stacking of wood, for example) may also support psychological wellbeing. Parents in Chapter 10 also found that the feeling that *others* perceive you to be competent important for wellbeing.

> The people from college walked through the coppice and saw what we have been doing in the coppice. People looked at my shelter and people talked while they walked through the coppice. They were amazed at what we had done in the coppice season and people commented on it [RMP03].

Student's echoing and sharing of ideas, behaviours and language surrounding the operation of the coppice with tutors demonstrate their understanding and absorption within its culture (Waite, 2013). It also provided many opportunities for their developing capacities to be confirmed, but also to connect with others through shared beliefs and outlook. The discrete environment of the coppice, built by students, separate from the college and only used by coppice workers, with its common understanding of the 'logic' of coppicing generated a cultureplace (Quay, 2017) within which students could share, mirror and relate positively with peers and tutors. The following examples show how the cultural norms of this place are taken on by students:

> [RMP11]: I don't like when somebody uses the tools in an unsafe way like when the axe, like swinging the axe too hard. Tutor: OK, you like to see a bit of respect. [RMP11]: Yeah…Tutor: Yeah. [RMP10]: It's like you [Tutor], axing that tree, that's the way it's meant to be done, like…. Tutor: To get into it with a bit of rhythm and [pause]. [RMP10]: Go fast.

> The next week after we first came here, we put the shelter, the tepee shelter up and we started work. We started tree felling, loping, sawing, axing, making shelters and taking wood to the pile. Putting wood on the fire and

> eating and drinking…in the coppice…Now it's the last week of the coppice and we're getting near to the end of the coppice season. The coppice season is from November to March. Spring Equinox. Winter Equinox to Spring Equinox [RMP03].

Students enjoyed participation and affirmation within the temporary intergenerational bubble of interpersonal relations, places and practices that emerged around coppice work. As they gained skills, they became part of a new community of practice (Waite & Pratt, 2015), which granted them access to a collective social and psychological identity—practical and productive managers of natural environment. This case study highlighted how context-dependent social, behavioural and environmental norms, outside of everyday expectations acquired through repeated group-based woodland activity supported psychological wellbeing (Fig. 7.2).

Top 3 messages from Ruskin Mill College coppicing with young adults with learning difficulties

1. For students, doing a 'proper job' with tangible lasting effects is key to their sense of achievement and purpose is important for psychological wellbeing. Consideration should be given to the purpose of activities and how this might contribute to different forms of wellbeing.
2. Sustained involvement enables the establishment of a distinct culture-place (Quay, 2017) that can build a supportive cultural density of new expectations of behaviour and competence (Waite, 2013).
3. Opportunities for progression in use of tools and responsibilities offer the chance for the development of growth mindsets (Dweck, 2008) where effort and persistence are valued.

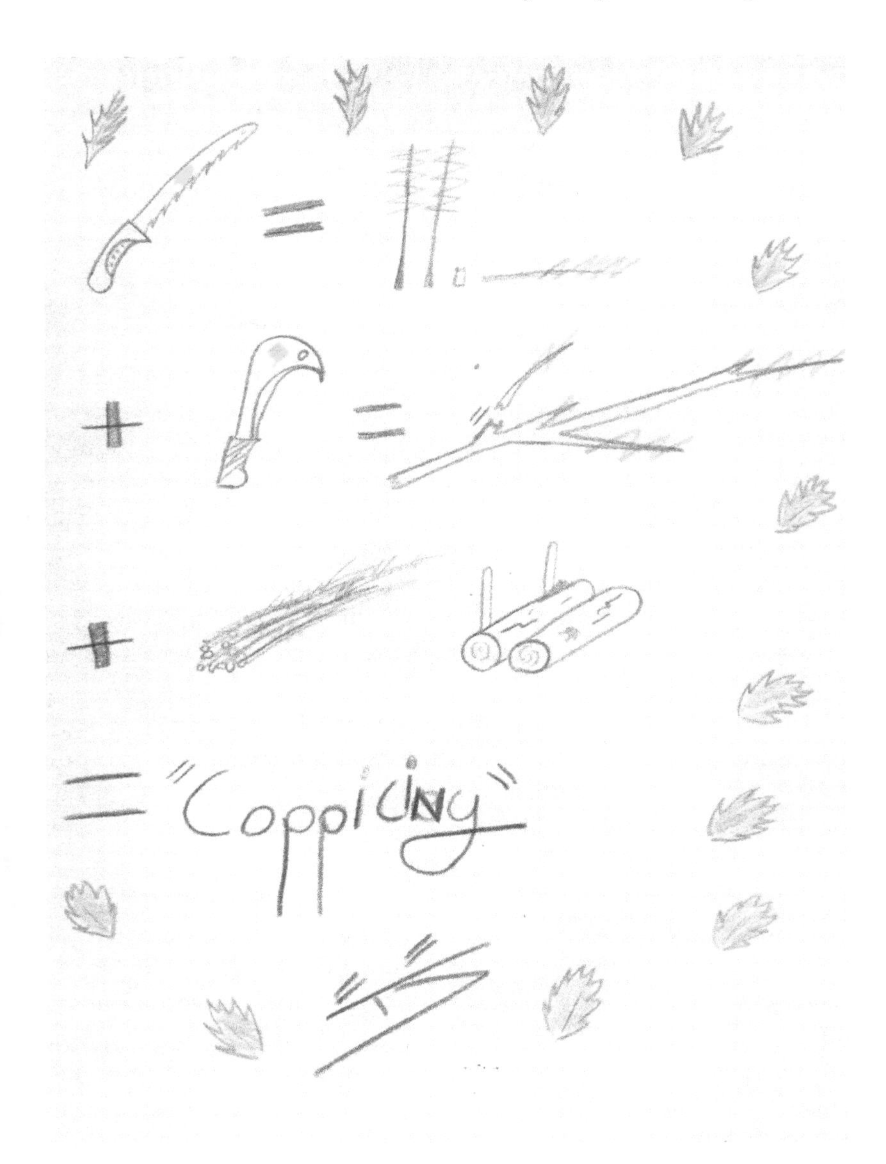

Fig. 7.2 Student drawing of coppice sums drawn during data collection (*Source* Ruskin Mill Local Partner's research evidence)

Case Study 2: Women's Wellbeing Through Woodland

The second case study, Timber Girls, concerns a project that drew on the approach and findings of the original GfW programme.

The Research Context

The Timber Girls project, funded by the Heritage Lottery Fund, was inspired by an opportunity to both celebrate the centenary of the Women's Forestry Service during World War I and a contemporary interest in why so few girls enter a career in forestry. In the UK, there is a significant gender imbalance within forestry and findings from the industry suggest that for girls a childhood love of the outdoors translates much less obviously into a clear career path, than for boys. Timber Girls aimed to examine with young girls and women their feelings towards woodland activity and forestry, and create a social, cultural and material environment that might help them challenge any preconceptions.

A workshop was designed to overcome barriers to girls taking part in forestry linked to a lack of exposure to forestry practices, and perceptions that it is primarily a masculine career. Girls came to woodland and were taught by female tutors to fell a tree with hand tools, crosscut the timber and move it with horses as it would have been done in past times. During the process, tutors shared some of the historical context for female forestry emerging during World War I, examples of its achievements and the fact that women's involvement in woodland management is very little remembered. A temporary cultureplace (Quay, 2017), womenworking-woodland if you like, was created through the setting (woodland), gender challenging behaviours (female tutors tutoring forestry a male-dominated industry) and cultural practices (historic references to forestry practices and female forestry). Womenworkingwoodland was largely unfamiliar to the young women, thus initially culturally light but it drew on long-established traditions and a cultural density from the past (Waite, 2013) and so was supportive of the girls taking on new roles and behaviours that were shaped by the way things had been done years ago. Thus, it was intended to be a context for changes in ideas, attitudes and behaviours for the young people.

During data collection and analysis, the researcher, Alice Goodenough, drawing on GfW experience and frameworks, also identified the impact of participation on student's happiness and satisfaction, alongside their description of gender-based barriers to accessing woodland wellbeing.

The Organisation

Courage Copse Creatives (CCC) is a small woodland enterprise based in the Taw Valley, North Devon run by owner/managers Katy Lee and Vince Large. CCC specialise in producing woodland products from their coppice that are both sustainable and ethical, including barbecue charcoal and pure biochar. They use low impact management methods including timber extraction with horses. CCC also run and host courses and workshops for both children and adults that are focused on promoting sustainable woodland management and use. Their educational and arts-based provision aims to achieve this by promoting sustainable engagement with and management of trees, woods and forests to adults and young people, but specifically to young women and girls who are under-represented in the woodland sector.

The Woodland Site

Courage Copse is a small 15.5-acre parcel of conifer plantation on an Ancient Woodland Site. In 1880, the broadleaf hazel coppice with oak standards on site was cut down and replaced with fast growing mixed conifer species. CCC are aiming to return the woodland to its earlier state by bringing the neglected coppice back into rotation and gradually removing the conifer and replacing it with native species mostly harvested from the woodland itself. The site is on a valley side that is steep in places, but the project activity took place in a flatter area of land being returned to coppice, where active felling of conifer was already taking place. The participants were involved in the felling of larch to both promote the coppice and to mitigate the risk of disease from phytophthora ramorum.

The larch was in poor condition, due to lack of thinning by previous owners. There was extremely limited biodiversity in terms of ground flora or animal species under the larch plantation. As a result of the tree felling during the Timber Girls project however, there has been an explosion of understory, return of wildflowers and an increase in insect life.

Forty-six girls (from North Devon schools and colleges) took part, the majority 13–14 years old, with a smaller cohort aged 17–18. They had volunteered to take part in the Timber Girls project as part of their school's or college's extra-curricular activity, frequently motivated by a pre-existing interest in spending time outdoors or in horses. Coming from a rural area of the UK, many had relatives or family friends who worked in agriculture, conservation or with animals. However, only one had a family member who worked in forestry (her father) and several knew people in professions they associated with forestry such as tree surgeons or forest educators.

Data was collected using four methods (with the informed consent of participating institutions and participants themselves): pre- and post-participation drawing survey; vox pop discussions at workshops to explore young people's experiences of the day; participant observation to record how young people responded to the workshop; documentary filming of workshops, for use post-project, as a prompt to further reflection/feedback amongst students, and sharing with stakeholders and interested audiences. Prior to the start of the project girls were surveyed for their perceptions of forestry through drawing a 'forester' and naming their key skills and traits. The majority imagined Foresters to be men (43%), rather than female (9%) or unisex (30%) (7% drew both male and female figures and 11% did not draw a figure), with physical strength and height their most notable characteristics. Timber felling was most often imagined to be what foresters did and some imagined that they were probably more courageous and out of the ordinary than other people.

In the following sections, we report some of the main ways that the girls reported changes to their psychological wellbeing and feelings of competence in this new setting.

Gaining Control, Challenging Norms

Learning and undertaking forestry skills with hand tools encouraged participants to share both how it made them feel, and to speculate why girls in general might not feel able to undertake such activity as a route towards greater wellbeing.

Girls frequently expressed feelings of inadequacy or caution at the start of practical activity, such as learning axing or sawing. However, with practice these tended to develop into more positive expressions of pleasure and positive regard for both others' and their own growing competency and control. Moving from...

> I feel like I'm going to fail...I'm going to miss it [the cut] ... [Peer] can do it...I can't...I'm scared I'm going to whack [her] [BR1 Year Nine Student]!
>
> This is actually rather dangerous; I'm surprised they let us go ahead with it [T Year Nine Student]?
>
> Do you just do it any way? How do I start? [T1 Year Nine Student]?
>
> I'm worried] cos it's sharp [BR5 Year Nine Student]!
>
> It feels a bit lame like, like ugggh...I've got no muscle [BC1 Year Twelve Student]!
>
> Getting there...my energy's been taken up by the axe...it's weakening my arms [B3 Year Nine Student].
>
> To...
>
> Yeah man, I love sawing [BR Year Nine Student]!
>
> Axing is so fun [T3 Year Nine Student]!
>
> Feel powerful [using axe] [Year Nine Group Discussion, T 25]!
>
> It's a nice noise [axe]...I love the smell of the [cut] wood...it's really satisfying [BC2 Year Twelve Student].

> Look how beautiful it is [log, cut and processed from tree] [BR Year Nine Student].
>
> You're being so neat [T3 Year Nine Student]!
>
> You're good at this…you're like making progress…[later]…that's well good [BC1 Year Twelve Student].
>
> Wow, you're doing great [BR1 Year Nine Student]!
>
> You're like worryingly good at this, like how many axes have you thrown at people?! [T5 Year Nine Student]

Girls expressed feelings may well contribute to psychological wellbeing: control, competency, developing oneself and being witnessed as competent by others, for example. It's perhaps to be expected that they sometimes felt self-conscious and unsure about learning new skills, experienced challenges in growing their ability and felt good when they overcame these. However, students taking part in these activities reflected that in fact considerable barriers might prevent girls feeling happy to undertake activity in woodlands (and therefore stumbling blocks in their access to the psychological wellbeing such activity supports). Though conscious that they might be stereotyping, girls suggested that in order to be confident to take part you would need to like the 'outside', 'mud', getting 'dirty' and so it 'Depends what kind of girl' [liv] you are (Fig. 7.3):

> …some girls are just like 'Yeah, we're girls I can't do that or something…', 'Oh my god It's mud, we're going to get dirty!' [Year Nine Group Discussion, T 23].
>
> I think some of them [peers] are just scared of manual labour really and some of them would be too worried about like breaking a nail [Year Nine Group Discussion, T 22].
>
> Like we're not the popular ones…like it there was a, like one of the popular kids here then it would just be a bit like…
>
> Awkward…they probably wouldn't want to do work to be honest.
>
> No, they'd probably, it would just be like…

Fig. 7.3 Moving timber with horses during Timber Girls project (*Source* and *copyright* The Timber Girls Project 2017, Courage Copse Creatives in partnership with North Devon Biosphere Foundation, funded by Heritage Lottery Fund)

> No-thanks [laughs].
>
> Ain't gettin' muddy [said in sing song tone]!
>
> ...I love getting muddy!
>
> Well I'm kind of used to it [laughs] [Year Nine Group Discussion, BR 09].

Getting or loving being muddy and liking the outdoors were sometimes described as interfering with the achievement of certain kinds of femininity and therefore regarded as obstacles to young women's engagement. Some girls attending the Timber Girls workshops felt they had already decided not to value such social and cultural objectives and were therefore able to participate and enjoy involvement. However, the students also repeatedly suggested that had boys attended the workshops with them then, regardless of their attitudes towards femininity, girls might not have been able to enjoy learning and experienced senses of increased capacity, control and self-development.

> They would just try and do it themselves, so then they'd start shouting at us, going 'no way, you're not doing it.'
>
> Or just start shouting at you 'oh no, you're doing it wrong.'
>
> When in fact they're doing it wrong and we're doing it right.
>
> 'No, [pretending to address male peers] YOU'RE doing it wrong' [BR1 & BR3 Year Nine Students]!

> They [boy classmates] wouldn't let us do anything.
>
> Yeah, they'd probably just take over, especially the boys in my learning group would never let me do anything [laughs].
>
> Because they always think they're better…in everything….PE they won't let me play because I'm a girl [Year Nine Group Discussion, BR 11].

> I think they'd take over and yeah, they'd yeah, like you [peer] said not take it very seriously and I don't think we'd get as much chance to actually have a go and sort show, either that or like some of the girls would be embarrassed to be actually doing it round boys.
>
> …I think they'd think that it's de-feminizing or something like that?
>
> Yeah and the boys would probably think it was de-masculating if a woman or a girl can do it better than they can [Year Nine Group Discussion, T 22].

> [Boys might] take the mick out of us. They'd be like "you can't wield a" what's it called?
>
> 'Oh what are you dressed like?!' and stuff like that [Year Nine Group Discussion, T 25].

Their comments suggest that this gendering and domination of masculinity was not confined to the traditional arena of forestry but was part of their everyday experience in schooling too, which some resisted. Timber Girls project not only raised awareness of male-dominated career paths, but the protected gendered space of womenworkingwoodland enabled them to express their frustrations in their regular experience as young women and the barriers these posed to realising aspects of their psychological wellbeing in an outdoor, practical context. However, their identification of what sort of women would choose or reject this cultureplace and how that had impacted participation within the project revealed that the status quo of gender relations would require further challenge.

Models of Alternative Being

It is difficult to trace how the sharing of role models from the past, stories of women from the Women's Forestry Service during World War I, contributed, although comments recognised the commonalities and contrasts with the current day.

> It has surprised me what they [female foresters] did in World War 1 and that. [Later]…it's nice to find out that people actually did do this and to do it is quite nice [student c]

> Even things all the way back there [gender equality], not actually that different to how they are today, you know, and doing the work today, not only has it been fun, but it's been interesting knowing what they would have actually done all those years ago [student a] ..

> It's good to know how far it's [gender equality] come, but it hasn't really come that far…it's still pretty backwards, like there's still huge, pay, wage gaps and women are kind of seen as the weaker sex [student b].

Opinions of what a forester was had changed over the experience, whether through consideration of womens' history, through the role modelling of women tutors doing woodland management or through the space for reflection that their active engagement in this new culturespace afforded. Participants' follow-up drawings and descriptions of 'what is a forester' changed in ways that suggest that positive interpersonal interactions with and observations of female tutors and consideration of women's contributions during World War I challenged contemporary barriers to female forestry. Workshops had encouraged girls to envisage foresters as female (Fig. 7.4).

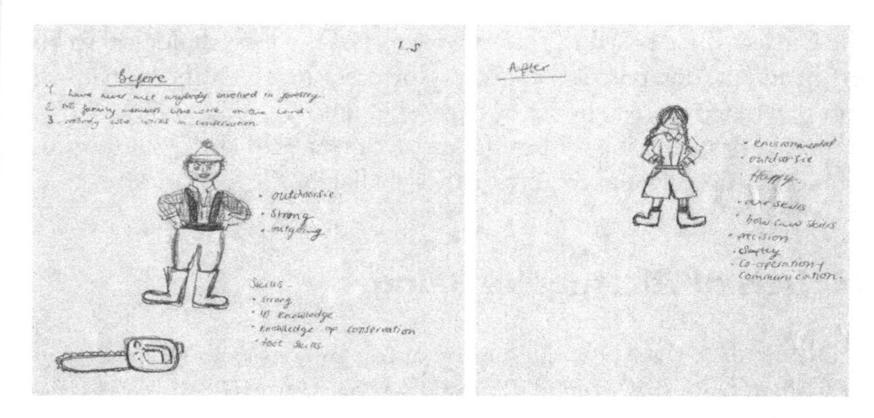

Fig. 7.4 'Before' and 'After' Timber Girls Project. Draw-A-Forester-Test survey example, student aged 13–14 (*Source* and *Copyright* The Timber Girls Project 2017, Courage Copse Creatives in partnership with North Devon Biosphere Foundation, funded by Heritage Lottery Fund)

Female foresters were drawn 33% more of the time (41% of drawings female), and no individual male figures were depicted (a decrease of 43%). There was an 11% rise in drawings that included male *and* female drawings of foresters (18% of drawings male and female). The number of unisex figures drawn (with no definitive markers of gender) remained constant before and after workshops (28%), (no drawing remained constant at 13%). Workshops also supported participants to focus less on masculine attributes or behaviours, such as exceptional strength and height and more on gender-neutral characteristics such as being hardworking or skilled. Participation also helped students to view foresters as 'normal', that is not needing unusual qualities such as particular courage. However, their experience reinforced and augmented their views that foresters must like the outdoors! All these shifts might contribute to encouraging female students to envisage themselves as able to participate in practical woodland activity and access support for psychological wellbeing through such behaviour.

No follow-up data collection was possible after the project end, so it is not possible to comment on whether the awareness of gender or psychological benefits of outdoor, woodland-based activity that the experience facilitated had longer-term effects back in school and in their daily lives.

Top 3 messages from Timber Girls

1. Forestry is often seen as a masculine activity requiring exceptional strength, and the industry is predominately male, but all-female participation in woodland management can help shift stereotyped assumptions and create a space for resistance to gendered norms.
2. Woodlands are however still seen as only for those who like being outside. Consideration of how to gradually expose those resistant to the outdoors is vital to open participation and the associated wellbeing benefits beyond the already converted.
3. The girls only practical activity context seemed as important in changing views and behaviour, alongside reflections on women's historical contributions to forestry. Active participation is an effective way to shake up thinking.

Evidence from these case studies surfaced the way in which physical discomfort or social and cultural norms may challenge engagement in woodland activity initially and therefore block access to psychological and other forms of wellbeing. Yet, the evidence also suggests that an element of challenge was important to create feelings of competency and achievement that were instrumental in psychological wellbeing. Immersion over a period of time in temporary woodland cultureplaces—spaces, social settings and behaviours (Quay, 2017) created metaphorical spaces that helped people tolerate or overcome this tension and experience psychological wellbeing and sense of functioning well. Other GfW case studies demonstrated that adults, as well as young people, can also experience psychological wellbeing through purposeful activity in which they gain competence. This can be either in relation to specific woodland skills or in the case of the National Trust families in the forest project through a sense that their engagement meant that they were being 'good parents' (Goodenough, Waite & Bartlett, 2015). A further point to keep in mind in designing programmes is that challenge supports the development of psychological wellbeing, but this only applies if the participants are not too vulnerable to meet those challenges (Sonntag-Östrom et al., 2015). At the least, it seems common for a period of negative affect to precede any subsequent elation at overcoming problems and that time and stepping stones towards a sense of achievement is a fruitful way to foster psychological woodland wellbeing across more diverse populations.

References

Dweck, C. (2008). *Mindset: The new psychology of success.* New York: Ballantine Books.

Goodenough, A., Waite, S., & Bartlett, J. (2015). Families in the forest: Guilt trips, bonding moments and potential springboards. *Annals of Leisure Research, 18*(3), 377–396. http://www.tandfonline.com/doi/full/10.1080/11745398.2015.1059769.

Herzog, T. R., & Strevey, S. J. (2008). Contact with nature, sense of humor and psychological wellbeing. *Environment and Behavior, 40*(6), 747–776.

Peschardt, K. K., & Stigsdotter, U. K. (2013). Associations between park characteristics and perceived restorativeness of small public urban green spaces. *Landscape and Urban Planning, 112,* 26–39.

Quay, J. (2017). From human–nature to cultureplace in education via an exploration of unity and relation in the work of Peirce and Dewey. *Studies in Philosophy and Education, 36*(4), 463–476.

Sanesi, G., Lafortezza, R., Bonnes, M., & Carrus, G. (2006). Comparison of two different approaches for assessing the psychological and social dimensions of green spaces. *Urban Forestry and Urban Greening, 5,* 121–129.

Seymour, L. (2003). *Nature and psychological wellbeing* (Report No. 533). London: English Nature.

Sobko, T., Jia, Z., & Brown, G. (2018). Measuring connectedness to nature in preschool children in an urban setting and its relation to psychological functioning. *PLoS One, 13*(11), e0207057. https://doi.org/10.1371/journal.pone.0207057.

Sonntag-Östrom, E., Stenlund, T., Nordin, M., Lundell, Y., Ahlgren, C., Fjellmann-Wiklund, A., … Dolling, A. (2015). "Nature's effect on my mind"—Patients' qualitative experiences of a forest rehabilitation programme. *Urban Forestry and Urban Greening, 14,* 607–614.

Stigsdotter, U. K., Corazon, S. S., Sidenius, U., Refshauge, A. D., & Grahn, P. (2017). Forest design for human health promotion—Using perceived sensory dimensions to elicit restorative responses. *Landscape and Urban Planning, 160,* 1–15.

Stigsdotter, U. K., & Grahn, P. (2011). Stressed individuals' preferences for activities and environmental characteristics in green spaces. *Urban Forestry and Urban Greening, 10*, 295–304.
Tyrväinen, L., Ojala, A., Korpela, K., Lanki, T., Tsunetsugo, Y., & Kagawa, T. (2013). The influence of urban green environments on stress relief measures: A field experiment. *Journal of Environmental Psychology, 38*, 1–9.
Von Lindern, E. (2015). Setting-dependent constraints on human restoration while visiting a wilderness park. *Journal of Outdoor Recreation and Tourism, 10*, 29–37.
Waite, S. (2013). 'Knowing your place in the world': How place and culture support and obstruct educational aims. *Cambridge Journal of Education, 43*(4), 413–434. https://doi.org/10.1080/0305764X.2013.792787.
Waite, S. (2015). Culture clash and concord: Supporting early learning outdoors in the UK. In H. Prince, K. Henderson, & B. Humberstone (Eds.), *International handbook of outdoor studies* (pp. 103–112). London: Routledge.
Waite, S., & Pratt, N. (2015). Situated learning (learning in situ). In J. D. Wright (Editor-in-chief), *International encyclopaedia for the social and behavioural sciences* (2nd ed., vol. 22, p. 5012). Oxford: Elsevier. ISBN: 9780080970868.
Zhang, J. W., Howell, R. T., & Iyer, R. (2014). Engagement with natural beauty moderates the positive relation between nature connectedness and psychological wellbeing. *Journal of Environmental Psychology, 38*, 55–63.

8

Natural Sources of Physical Wellbeing

What Is Physical Wellbeing?

Health is often coupled with wellbeing in current discourse about public health, in line with the old adage of 'a healthy mind in a healthy body'. However, this holistic view has not always been the case in clinical treatment of illnesses and the current coupling also recognises a shift in medical approaches towards more preventative forms of treatment. A positive approach acknowledges that promotion and maintenance of health and wellbeing will avoid or ameliorate illnesses and that wellbeing itself should not be viewed as simply an absence of illness, but as an entitlement to a good life for all. Physical wellbeing is intimately connected to other forms of wellbeing but is principally derived from lower levels of sedentary behaviour and more physical activity with implications for corporeal as well as mental health. The Good from Woods Project pilot framework identified physical wellbeing as when people in woodland were feeling confidence in and enjoyment of physical activity.

A. Goodenough and S. Waite, *Wellbeing from Woodland*,
https://doi.org/10.1007/978-3-030-32629-6_8

Moving More and Exposed to Nature

There is a strong and growing body of evidence that regular physical activity in natural environments has physical and psychological benefits (Barton, Bragg, Wood, & Pretty, 2016). However, the causal relationships between elements within green physical activity and 'green exercise' (a term coined by Pretty, Peacock, Sellens, & Griffin, 2005) are less well established. Mitchell (2013), for example, cautions that physical activity itself is linked to mental health (Hamer, Stamatakis, & Steptoe, 2009) and that certain environments might simply stimulate higher levels of activity than others, rather than effects being associated with the natural context per se. On the other hand, there are studies that show even passively viewing a green space can enhance wellbeing and recovery rates (Ottoson & Grahn, 2005; Raanaas, Evensen, Rich, Sjostrom, & Patil, 2011; Ulrich, 1984) as explored in Chapter 1 and that natural context has a contribution to health and wellbeing in itself without physical activity. It is worthwhile trying to explore some of the complexity of the effects of nature and activity and how the interaction of the two may multiply benefits (Rogerson & Barton, 2015), so that programmes aimed at capitalising on natural resources for health and wellbeing can be tailored appropriately.

Theories put forward to explain possible motivations for physical activity include Social Cognitive Theory, the Theory of Planned Behaviour, Self-Determination Theory and the Transtheoretical Model (Sallis & Glanz, 2006). They focus on intrapersonal factors with little attention paid to how environment influences behaviour and potential health outcomes.

As discussed in Chapters 1 and 2, a number of explanations focused on evolved adaptations suggest ways in which the environment may support physical activity and green exercise. The biophilia hypothesis claims we are programmed in evolutionary terms to prefer being in nature (Kellert & Wilson, 1993) and are therefore motivated to spend time there. Complementary explanations, ART (Kaplan & Kaplan, 1989) and PET (Ulrich, 1981) emphasise how physiological changes in body and brain associated with the restorative effects of exposure to nature may support physical health and exertion: including boosted immunity, reductions in blood pressure, stress hormones and heart rate, enhanced directed attention, improved self-awareness and self-esteem (Pasanen, Ojala, Tyrväinen, & Korpela, 2018; Pryor, Townsend, Maller, & Field, 2006).

These theories also essentially focus on *within*-person factors from a psychological perspective. However, others (Brymer & Davids, 2012, 2014; Mitchell, 2013; Yeh et al., 2016) have suggested multilevel ecological, dynamic approaches are needed to take account of the fluid relationships between environment, activity and individual during green physical activity. Yeh et al. (2016) note that any of these three factors may constrain or invite physical activity, but being active in *natural environments* (in comparison with indoor settings) requires us to negotiate 'changing olfactory, acoustic, haptic and visual information' which may all influence physical wellbeing (Yeh et al., 2016: 4).

> one can feel wind, sunlight, rain...distinct textures, terrains and surfaces...-sounds from birds, water or smell from flowers, trees...feedback from the plantar surface of the feet while walking/running/stepping...or the hands while climbing which invite richer psychological responses than when undertaking PA [physical activity] in more static conditions of temperature-controlled, enclosed gymnasia (ibid).

Physical activity in natural and indoor environments may also be informed by our cultural understandings of and expectations for behaviour in that setting. Quay's (2017) 'cultureplace' approach, described in previous chapters, would suggest further flow between environments and person towards a unified melding of individual meanings with particular places. Every place holds a different significance for individuals, but as Waite (2013) argues, this is highly determined by established social practices associated with them—the 'cultural density' of places. This complex interplay argues for careful attention to nuances in environment, activity and individual and a recognition that one-size-fits-all approaches that do not account for sociocultural effects are likely to be flawed.

Pasanen et al. (2018) point out that much research about associations between wellbeing and nature is laboratory based. Along with Markevych et al. (2017), they also suggest that there are other pathways/influences to consider in an ecological approach to understanding physical activity, alongside psychological restoration (environmental impacting on mental).

These include reduction of harm from pollution and noise (environmental impacting on physical) and capacity building (cultural impacting on social through physical and mental). Their study based on recall of physical activity in different contexts found that context alone could not explain differences in restoration and wellbeing and that activity choices and social factors also impacted, in line with other studies such as White, Pahl, Ashbullby, Herbert, and Depledge (2013).

Using self-report tests of health and wellbeing, Mitchell (2013) found that regular use (at least once a week) of open space/park or woods/forest areas for physical activity was associated with a substantially lower risk of high scores on the General Health Questionnaire (GHQ12), compared with those not using these environments. No significant association was observed for any of the other environments considered, which included sports facilities, beaches, gardens and footpaths. Effect sizes were also notably larger for the woods/forest category when used regularly twice a week or more. For those using woodland for exercise twice a week or more, the odds of having a lower GH12 score being due to chance were less than 0.05. However, the results were complex, in that regular indoor physical activity was associated with higher Warwick Edinburgh Mental Health and Wellbeing scores. Since GHQ12 measures more severe health problems, it may be that the indoor activity is better suited for those with milder mental health issues. Reinterpreting this from a cultural perspective, it may be that sports centres are more conducive cultureplaces (Quay, 2017) for those who are already experiencing better health. For those who are not sporty or feeling well, such places may clash with their usual habits, abilities and dispositions, be off-putting and counterproductive to changes in behaviour (Waite, 2015). This might indicate that protecting, developing and promoting woodlands and parks could be particularly useful in addressing health outcomes for people with more serious health issues, especially as choosing to exercise in these places just once a week appeared enough to achieve benefits.

While green exercise has many evidenced benefits, the relationship and pathways to health and wellbeing are not yet fully understood, but embodied and active engagement with nature clearly has a role for some people in boosting physical wellbeing. In this chapter, we look at several Good from Woods case studies that focused on the physical and moving more amongst trees.

Case Study 1: Learning on the Move

The first case study concerns a partnership project between Mayflower Community Academy, Plymouth City Council's Natural Infrastructure Team and the Friends of Ham Woods. The Mayflower Academy was participating in the Natural Connections Demonstration Project (Waite, Passy, Gilchrist, Hunt, & Blackwell, 2016) which aimed to embed curricular learning in local natural environments (LINE), and the partnership was keen to establish some of the effects of the Academy's LINE provision and the role of local woodland as a green learning environment. The result was a qualitative study designed to explore the physical health benefits and wellbeing outcomes associated with the children's weekly outdoor learning in local woodland. Ten school children took part aged 6–7.

The practitioner-researchers were from different backgrounds; Jennie Aronsson was a school health nurse and Naomi Wright, an environmental scientist and artist. Jennie had carried out an earlier study 'Woodland Health for Youth (WHY)' of the children's physical activity using accelerometers which the qualitative study was intended to complement (Aronsson, Waite, & Tighe-Clark, 2015). The WHY study was inspired by the risk to children's health of increasingly sedentary lifestyles and high-calorie diets (Health and Social Care Information Centre, 2015). Preventing childhood obesity through raising activity levels is a national priority (Public Health England [PHE]) that can also help to address muscular strength, bone health, cardiorespiratory fitness, self-esteem, anxiety/stress, academic achievement, cognitive functioning, attention/concentration, confidence and peer friendship (PHE, 2018). Only two in ten children aged 5–15 years meet the national physical activity target of 60 minutes or more moderate-to-vigorous physical activity (MVPA) every day (PHE, 2014) with children from the lowest income bracket more likely to report low levels of activity (PHE, 2014). School-based initiatives to increase physical activity offer a universal means to reach diverse populations of children (Kriemler et al., 2011), but stand-alone health promotion initiatives can clash with other priorities in schools. The WHY project was intended to evaluate whether outdoor learning would allow for equitable access to physical activity interventions for school-age children without compromising school's delivery of core subjects.

The Woodland Site

Ham Woods is a public local nature reserve of over 35 hectares within the city of Plymouth. With a brook and paths running through it, it has a mixture of open areas and more wooded parts. The site has more than 200 species of plants and almost 80 species of birds (Plymouth City Council, 2014). Amongst regrowth of ash and sycamore and large older 'marker' trees, there are some non-native specimen trees, probably planted by the owners of the old manor, Ham House. There are other cultural artefacts such as a historic leat to a local mill, World War II bomb craters, old lanes and stone walls, adding to the interest and quality of the valley (Friends of Ham Woods, 2014). Most of the LINE activities took place in hollows and valley 'cross-roads' about 20 minutes' walk from the school. A favoured spot was a circle of logs around a fire pit zone. Another was a square of dry ground surrounded by trees and bordered by the stream. One which featured a 'climbing' tree.

The woodland was in a neglected state prior to Plymouth City Council's Natural Infrastructure Team work with the local community through a lottery-funded programme, Stepping Stones to Nature. Ham Woods was then maintained and promoted as a community resource under the auspices of the Friends' group. Accessibility in and around the woodland was improved by creating the paths, designing signposts and repairing bridges. An initial programme of community engagement through walks, litter picking and family events was organised by the Stepping Stones to Nature team, and Friends of Ham Woods then took up the challenge to run a range of activities by and for their community, such as bug hunts, woodcraft and play-days, to promote their continuing use and conservation.

Mayflower Community Academy had been providing Forest School sessions in Ham Woods since the school opened in 2009 but since September 2013, as part of Natural Connections, it had extended this to curricular experiential learning. While the primary focus of sessions in the woods was educational, the school visits to the woods also involved physical exercise, including the walk to the site and many learning activities themselves involved physical activity. The LINE Facilitator at Mayflower Community Academy was a trained Forest School leader. She or another teacher would take the children outside for learning in the natural environment (LINE) one afternoon a week regardless of the weather, supported by a teaching assistant and a volunteer from the Friends' group. The practitioner-researchers also participated in activities during sessions they attended (Fig. 8.1).

Jennie and Naomi used several methods to generate data and found conversational drift (also discussed in Chapter 9) particularly useful in adapting to the changing needs and priorities of the participants. This is a term borrowed from artists working in this way (Adcock, 1992) to describe the approach of using conversation in the woods and letting the conversation take its own path. As opposed to discussion groups, it allowed the main activities to continue without impinging on curriculum time or enjoyment. They found that in order to engage the children aged 6–7 years, more playful methods such as this, within the context of a perhaps more playable environment, were more successful. Games were devised to capture children's views. Interviews worked well with the adults who were better able to talk at times other than during the outdoor sessions. A variety of visual methods such as videos, photography and observational drawing

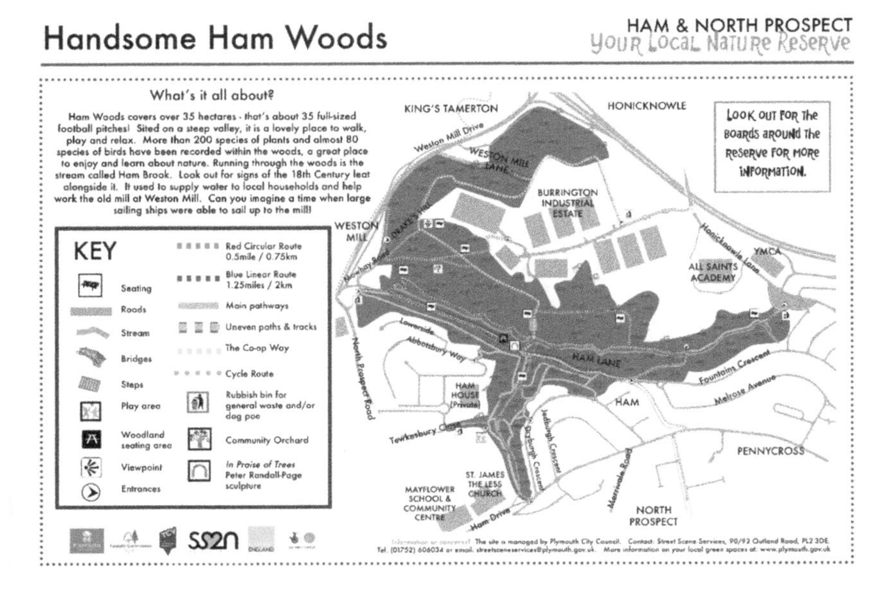

Fig. 8.1 Map of Ham Woods site and school location (*Source* Local partners report)

also tapped into different aspects of the experience and were especially valuable for the physical aspect of wellbeing. Observational drawings were able to capture movement through the area and engagements with material features in the landscape in a way that children of this age would find difficult to articulate. The two researchers also kept a reflective journal which aided interpretation of data. They noted and agreed the presence of different indicators of forms of wellbeing, and in this chapter, we focus on those related to physical wellbeing, which was the most dominant theme in the data.

It was interesting that physical wellbeing indicators emerged significantly less frequently amongst adult understanding of outcomes (teachers, teaching assistants, parents, volunteers and a governor—33 times) than amongst the data from the children (144 times). While children couldn't stop themselves demonstrating or talking about activities, adults tended to mention after-effects of physical activity such as being 'happy tired'. For parents, children's enjoyment was paramount. School staff focused on educational benefits of LINE, while Friends of Ham Woods and the Natural Infrastructure Team emphasised the opportunity to connect children with nature. Physical wellbeing effects were more than twice as often noted as the next most common type, emotional wellbeing. While this was also influenced by the focus of this piece of research, the observed embodied actions and data from the WHY project provided corroboration of its importance to children.

The most common indicator of physical wellbeing was 'Doing a physical activity' that captured children's keenness to talk about or demonstrate their activity in woodland and how the cultural space of woodland gave them permission to do so. It was noted 63 times in the data. One child was noted to be less enthusiastic, but even for those pupils who were observed to less active than peers at break times, LINE experiences seemed to stimulate more movement. LINE was associated with children across the board becoming more active, and the affordance (the cues and invitations to use they suggest) of trees and wooded environments were implicated in this increase, as in the following examples:

Q: With friends and family, do you do the same sort of activities that you do with school?
C2: Ah, we play hide and seek.
C4: then run across so many trees, and then I jump down.
C1: When I go to my nan's, she's got a wood and I go up the woods and climb trees.
[...]
Q: Why do you like climbing trees then?
C4: Because sometimes I pretend that I'm a squirrel because I'm really careful in trees, I climb up there and I jump out the top.
Q: Have you got a favourite activity in the woods?
C4: Playing games, playing in the mud, slipping down, playing up in trees, and picking leaves.

Some children deliberately sought out opportunities to be active, and the physical activity was often prompted by the context in which activities took place.

'Confidence in and enjoyment of physical activity' was mentioned by children 48 times, the third most common indicator in total. The main contributors to pupils' confidence and enjoyment were experienced as cultureplaces such as goingintheriver, climbingtrees and slippinginthemud, where children's activity and material woodland environment formed a unified event. Interestingly, physical activity in the woods could be experienced as energy producing.

Q: Which do you like best, running or walking?
C8: Running.
Q: Why is that then?
C8: Gives you more energy.

Although running is in fact energy consuming, it releases endorphins that can create a feeling of euphoria and energy.

The practitioner-researchers also noted the importance of imagination within woodland wellbeing in this case study. For some children it seemed place, their creative imaginings and activity were the dynamic constraints to and invitations for green physical activity. The wooded environment inspired imagination shaping action.

Q: How does it make you feel if you are looking at the trees now, as you are now, how do they make you feel compared to sort of being in school?

C6: Like I'm just about to go on a big adventure.

C4: I like crouching down because I think that there's bears looking for us.

From accelerometer data measuring the intensity of physical activity (Aronsson et al., 2015), we can see in Fig. 8.1 that the woodland environment stimulated more intense physical activity than other outdoor contexts. The figure shows the difference between the proportion of time children spent in moderate and vigorous physical activity (MVPA) depending on if they were engaged in woodland LINE, school grounds LINE or an indoor lesson. This concurs with previous studies which concluded that children who spend more time outside are more active (Cooper et al., 2010; Stone and Faulkner, 2014) (Fig. 8.2).

The analysis also captured both awareness (20 instances) and absence of awareness (11 instances) about physical health amongst pupils (under the indicator of wellbeing 'Feelings of physical health'). For example, one girl responded 'No' to the question: Do you ever get out of breath with

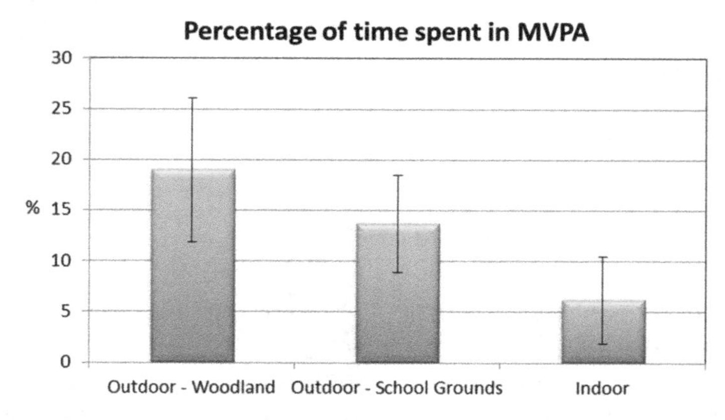

Fig. 8.2 Graph showing difference between the proportion of time children spent in moderate and vigorous physical activity (MVPA) depending on whether they were engaged in woodland LINE, school grounds LINE or an indoor lesson (*Source* Aronsson et al., 2015: 59)

all this moving around? Yet, observations showed that she was out of puff as she ran around but was simply unaware of this as she was engaged so deeply in her actions. Some children, however, mentioned getting hot and sweaty or getting out of breath as making them healthy. They linked this to their physical activity and being in the woods. It is not clear whether being aware of health outcomes is important to encourage more participation in exercise. It may be that being lost in enjoyment may be more motivating, with physical wellbeing outcomes a useful by-product. It may also suggest that there is a developmental aspect that children become more physically aware as they grow older. Other research including an older age group suggests that awareness could provide an extrinsic goal for children if they monitored their physical activity through pedometers and charted progress. A toolkit arising from this research to help schools promote health and wellbeing outcomes is available from the University of Plymouth's Peninsular Research in Outdoor Learning website (Wright, Waite, Graham, Aronsson, & Waite, 2016).

In the next case study, the participants were teenagers, which give us a useful opportunity to consider physical wellbeing indicators at different ages in childhood.

Case Study 2: Forest School for Young Teenagers

A GfW partnership project between Otterhead Forest School and The Castle School, Taunton aimed to explore any effects of regular Forest School sessions on young teenagers' wellbeing. Thirteen students who were experiencing difficulties in the classroom were chosen by the school to participate in a six-week day-long Forest School group. Some of the young people had also previously attended Otterhead's Forest School programmes. The practitioner-researcher, Jenny Archard, had a long history of working with children, young people and adults in Forest School and woodland activities. Her main focus for the research was to observe and document impacts of the Forest School process on young people, especially those who only came for six sessions. She was interested in identifying any areas for improvements in practice and contributing to the evidence-base around Forest School work beyond early years and primary aged children. She used interviews, observations with photos and videos with a reflective journal to aid interpretation of data. To track any changes in their attributions of wellbeing, Jenny carried out a follow-up study six months later in different woodlands with the support of the GfW project researcher. As an innovative method to help confirm what students thought they had got out of Forest School, the practitioner-researcher 'translated' the GfW indicators into 20 experiences she felt would be familiar to these participants. These were printed out as statements that participants could physically sort and order into what most contributed to them feeling good (who, what and where). Her subsequent analysis used the set of wellbeing domains and indicators of wellbeing being developed collaboratively by the research team and practitioner-researchers (Good from Woods, 2014) (Fig. 8.3).

The Woodland Site

Otterhead Forest School is based in an old coach house at Otterhead Lakes, a mixed woodland local nature reserve of some 230 acres in the Blackdown Hills Area of Outstanding Natural Beauty. Members of staff at the time of the research were pioneers in the use of the Forest School approach with older children and young people with behavioural issues. They worked mostly with young teenagers needing social, emotional and mental health support, either in groups or on a one-to-one basis.

The young people taking part in the research were aged between 11 and 16. They included five students aged 11–13, new to Forest School and coming a day a week in a six-week block and eight students aged between 11 and 16, who had already been attending for between six weeks and four years. All the students, to varying degrees, had emotional and behavioural needs. Sessions were led by two Otterhead Forest School staff, with the project researcher; although students arrived by school minibus, school staff did not stay with them.

Here, we focus on the physical wellbeing outcomes that emerged from this research with teenagers (elsewhere we have discussed other aspects of this case study, Waite & Goodenough, 2018). In preliminary research,

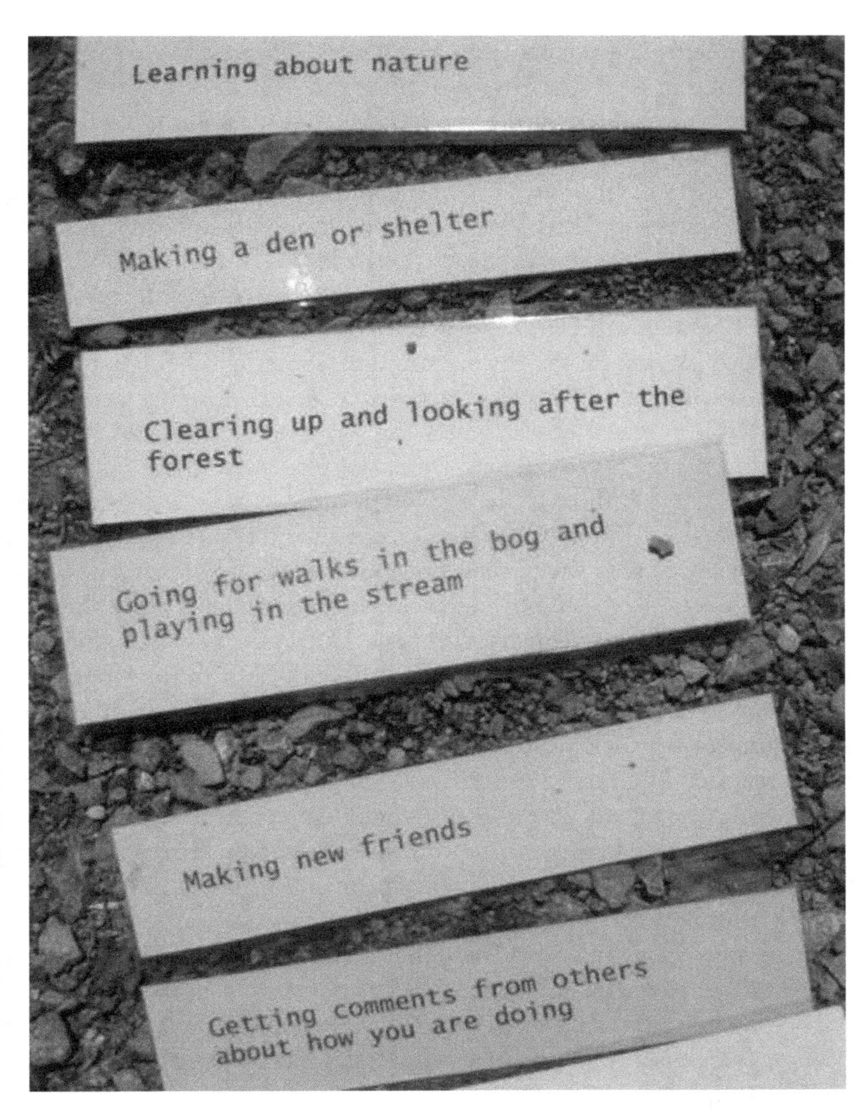

Fig. 8.3 What, where, who with sorting exercise used to test and confirm findings and establish long-term outcomes at Otterhead Forest School (*Source* Otterhead Forest School Local Partner's research evidence)

the school referring students and Otterhead staff emphasised social and emotional outcomes as their expectations from the Forest School experience, but few mentioned nature connection and physical activity themes. Preliminary conversations with new students found they expected to do things that were physical and fun and could potentially lead to learning about social, emotional and behavioural development, alongside practical skills gain and knowledge of the natural world.

During data collection and analysis, Jenny was struck by the extent to which most students did enjoy green physical activity: walking, climbing, running and playing on site during sessions. Sometimes being physically energetic was a 'free' activity that students created themselves, and sometimes it was supported by staff after the learners had initiated it. They also showed confidence in and enjoyment of physical activity. As soon as they got off the minibus, some would run down the hill, chasing each other and laughing. Students engaged in rough and tumble with staff and were physically close to them, sometimes even clingy. However, one of the most visible ways in which physical wellbeing was evidenced was through the young people's engagement with play. The students played the most while in the woods, but also played on the walk down to the woods and on the way back. Playing could happen on their own, in small groups, large groups, with sticks, with mud and with staff. It was sometimes as a game organised by students or staff, or it might be initiated informally by students. At other times, it appeared to bubble up completely unplanned.

What was especially interesting was the longevity of these experiences and that physicality was a dominant theme in enduring memories of wellbeing when followed up six months later. Biophilic indicators of 'feelings of closeness to the natural world' and 'Being engaged in a relationship with nature' also emerged as significant. These aspects of woodland wellbeing had neither been observed by the practitioner-researcher nor described by students frequently in the first round of data. However, six months later these two experiences (clearing up and looking after the forest; being in the woods) were most recalled by students as leading to wellbeing, perhaps indicating that they had become more important over time.

Jenny's findings can be seen to support a more dynamic understanding of green physical activity, one that recognises the interplay between environment, individual and activity. In this instance, play as a physical activity seems to have contributed as a pathway to young people feeling engaged with and close to the wooded environment in a way that was unanticipated. It is possible that the 'olfactory, acoustic, haptic and visual information' from the natural world that Yeh et al. (2016: 4) identify as influencing the psychological effects of green physical activity also provides young people with a sense of coming into relation with nature. 'Playing' in the wooded environment at Otterhead brought students closer to nature.

Q: …what about you, best bit [of Forest School] today?
A2: Getting my face covered in water.
A3: When he's jumped to [the] end, slipped down and it's sloshed up into his face, it was incredible.

Students distinguished between 'messing around' and more purposeful 'play' they pursued. Both messing around in the woods and being triggered into activity through its cues to imaginative play (as described in the case study above) brought students into a relationship with the natural, material environment.

A2: Mmm, it's not exactly messing around [with sticks], you [have made] a spelling error, its 'playing' around.
A1: Correction, 'playing with sticks', well actually there's a difference.
Q: So why is playing with sticks different?
A2: Because you play with them you go like—'I'm going to fight you with my sword' [mimes sword action].
A2: Whereas if you're messing around, you're going to go like this [mimes smacking action]
A1: You're going to smack people with it.
Q: Do you think everyone was playing or messing around with sticks earlier on?
A1: Messing around.
A2: Some were messing around and the ones that want to do sword fights.
Q: OK, so there's a difference between playing and messing around.
A2: Big difference.

This finding also suggests that wellbeing outcomes may occur on different timescales and that during outcome measurement, it is sometimes difficult to attribute what effects different practices have. As we noted in the previous case study about physical activity through curricular learning, some children were not aware at the time that their actions were impacting on their health and wellbeing, but this did not mean that these effects were not taking place. Our case study about planting trees and later environmental attitudes (Waite, Goodenough, Norris, & Puttick, 2016; Chapter 8) also corroborates an understanding of potential slow burn impacts as well as immediate effects of exposure to trees. It may be that much research looks for the 'wrong' indicators or measures at the wrong interval to judge if children and young people are making a connection to nature through relatively short-term interventions. Perhaps it is not always a longer intervention but a longer follow-up to assess change that is needed.

The differences between stakeholders' expectations and students were also an unexpected part of this research and have implications. Firstly, if Forest School is expected to be led by the participants rather than a previously set agenda, attention to what young people's expectations are is vital. Sharing clarity about these expectations would help to bridge gaps between Forest School and the institutions from which participants come. Play and physicality were dominant themes for the young people, which suggest service providers should ensure that play is not something that gets forgotten in Forest School for older age groups. Play is not only a potential pathway to positive emotions and nature connection but embodies autonomous action and therefore contributes to a sense of control and self-regulation (psychological wellbeing). The opportunity to play may be rare in secondary students' schooling experience. Further research could also explore the difference in wellbeing outcomes between young people in regular woodland sessions who are engaged in predetermined nature connection activities, and those whose leaders let those activities develop organically, exploring the balance between structure and freedom (Waite & Davis, 2007).

Case Study 3: Centre for Contemporary Arts in the Natural World (CCANW)

The Centre for Contemporary Arts in the Natural World (CCANW) carried out two research projects to look at impacts on wellbeing from two physical engagements with woodland: Forest Football and Dancing Trails. These were activities taking place as part of their 'Games People Play' exhibition and activities for the Cultural Olympiad running alongside the Olympic Games. CCANW were seeking to encourage people to engage creatively with ecological issues through the arts. The 'Games People Play' programme explored what games might show us about 'human nature' and how an understanding of what cooperation can achieve might support us in addressing planetary needs.

The Woodland Site

At the time of the research, CCANW was based at Haldon Forest near Exeter; it has since relocated to Dartington Hall, Totnes. Haldon Forest Park is owned by the Forestry Commission and comprises over 3500 acres of forest, mostly conifer plantation with Sitka spruce which is used for paper and in the building industry. There are also Scots pines, lodgepole pines, silver birch and Douglas fir. The forest was established in the 1920s and is now a Site of Special Scientific Interest (SSSI) for wildlife. The woodland has numerous trails for walking, jogging, mountain biking and leisure facilities such as 'Go Ape', an education/visitor centre and cafe. It organises events throughout the year for all ages, including mindfulness, Nordic walking and orienteering.

CCANW and their practitioner-researcher, Gemma Baal, were interested to find out wellbeing outcomes from Forest Football. Forest Football was a physical team-based activity where members of the public played a version of football in the woods. The goals were painted 'Mayan' hoops hanging from the trees, rather than nets on the ground. Rather than the usual delimited smooth grass field, the game ranged in and amongst the trees, rocks, roots and ploughed earth. Staff and the artist-designer expected that its unfamiliar forest setting might act as a 'leveller', helping people to drop cultural expectations of required behaviours such as being 'fit' and specific skills and hence be empowered to join in. The artist-designer also felt that the 'green' context and opportunities to engage sensorially with nature through the game could impact positively on players.

Gemma gathered 17 participant views through interviews and paper-based questionnaires and found that these aims had been achieved in encouraging those who would not normally engage with football getting involved.

> there's like a range of ages to just have like throwing the ball about and trying to get it in without focusing too much on the sport [and competitiveness] cos everyone is of different abilities (V1).

> it's kind of alternative sport [in the forest] which is a bit funner I think...things like sports halls get me a bit like, it just seems very sporty and I'm not that sporty (V2).

> 'I said to him (son), "are you having fun" he said, "yeah, it's really weird". He loves things being weird; he's not so keen on mainstream' (V4).

Making the sport unusual was found to be a key source of wellbeing benefits in line with expectations. It allowed participants who didn't usually play sport to feel that they stood a chance against more experienced footballers. The uneven forest floor and unusual goal layout meant that football experience didn't necessarily lend an advantage and the trees interrupted views of goals and other players making it a very different context to a traditional playing field. The outcomes were that people were inspired by the cultural freedom it inspired (Waite, 2013) to move actively in the forest. Forest Football was a clash with the culture and habits participants associated with football.

The second study focused on wellbeing outcomes for audience members achieved by taking modern dance out of a theatre setting and into the forest trails where people might be drawn into the performance and the forest. In this study, members of the public followed the dance through the forest, so were more physically active than a conventional seated audience. A specially composed music score and choreographed dance explored themes of fairy tales, discovery and walking, using props such as clothing, a ball of wool and a vintage gramophone. The audience wore wireless headphones, so they could hear the accompanying music. The dancers hoped that the narrative and movements of their performance could bring people into new perspectives on both the forest and the imagery and themes of the dance. The composer hoped that the music might 'isolate' the walkers from external influences taking them on an 'inner journey'.

Using the same methods of interviews and questionnaires with 15 audience members, Gemma found some conflicting themes on the physical experience of participants.

> Well I thought it was very good, yeah it was, I don't know it was really different, that's why I liked it, like normally you're just sitting there watching all these people just dancing around the stage – [here you] just follow them and interact with them, there's just natural obstacles, it was cool (A2).

> Interesting, some parts of [the experience] more so than others. Some I was able to recognise and associate with previous experiences but some of it was completely new, so curiosity won out. You know hiding in the bushes..., standing in the bushes...I felt more connected to the dancers...Doing that rather than being part of the group that's blocking the trail (A3).

> I took mine off quite a few times (headphones) so I could hear something particular, sometimes it's a bit irritating, you hear lots of loud noises, you want to experience the real forest as well (A4).

> I was in a beautiful forest for the first time I think .... so, I would have been happy just hearing the sounds, music and singing and laughter in one ear and the natural sound in the other ear (A5).

Although the novelty of the performance seemed stimulating for some, confusing the norms of both dance performance and forest use seemed a barrier for others. In particular, passive physical activity that seemed at odds with familiar natural world cues and invitations for behaviour (listening, looking) and human use (sharing pathways) seemed to distract from wellbeing. While these two case studies drew on disruption to the everyday and our usual habits, dispositions and skills (habitus) which was found to be effective in supporting wellbeing in other case studies in this book, it seemed to have mixed results in these instances (Waite, 2013). While the forest may be relatively culturally light for some people and open for new experience, cues to activity provided by nature or more familiar habits within it sometimes clashed with appreciation of the performance (Waite, 2015). Incongruence can be both helpful and obstructive, depending on the individual and circumstances.

Implications for Promotion of Physical Wellbeing—'Moving More' and 'More Moving'

These case studies have shown that being physically active in woodland offers opportunities to develop wellbeing. Allowing primary school children to move more while learning addresses concerns that some children struggle to concentrate when expected to keep still and listen rather than learn experientially (Waite, 2010; Waite, Rogers, & Evans, 2013). In line with current prioritisation of preventive medicine, it also helped reduce the amount of sedentary behaviour which has implications for children's weight and fitness. This is especially significant for children from socioeconomically deprived areas, as research shows that children living in poverty are nine times less likely to have access to green space, places to play outside and to live in environments with better air quality (National Children's Bureau, 2013). Child poverty levels are higher in Plymouth than in England overall; 20% of children under 16 in Plymouth live in poverty (Public Health England, 2018), and this research was conducted in the 3rd most deprived area of Plymouth (Plymouth Fairness Commission, 2013). However, the research showed that these socio-economic barriers to outdoor health promotion in green space could be overcome by offering outdoor learning as part of the curriculum so that all children regardless of background gained access to natural environments through their school. Furthermore, individual differences in levels of sedentary behaviour were equalised through this form of universal access. Children monitored for levels of physical activity using accelerometers showed that their physical activity was more evenly distributed during outdoor learning in woodland compared to outdoor learning in school grounds or in the classrooms or at break times. While at playtimes children tended to follow their customary sedentary or active behaviours, lessons that took place in woodland elevated all children's activity levels. This could be linked to the rich cues for imagination and physical activity provided by nature within this setting. Other research has suggested that the closer both geographically and culturally that the space is to the classroom, the more those institutional norms will be reproduced (Rogers, Waite, & Evans, 2017). It may be that woodland represents a 'wilder' culturally light environment to establish new ways of behaving.

The need to be active and continue to be playful was evident in teenagers too. At Otterhead Forest School, the freedom to move more reduced reported stress levels for students that were finding mainstream schooling difficult. In the longer term, nature's invitations to physicality and fun amongst the trees provided an engagement with the natural world that led to greater awareness of and care for the environment. That woodland can stimulate playfulness, movement and care, leading to enhanced wellbeing has clear implications for the design of places for youth to inspire such behaviours.

The relocation of sports and dance into the forest seemed to encourage those who would not usually take part in these activities. Physical wellbeing may partially stem from the opportunity to act differently from the norm, particularly if the norm is a setting in which you feel less comfortable. As we have discussed, offering this difference in contexts that encourage physical engagement can disrupt self-perceptions within dominant sociocultural frameworks of understanding ourselves and our abilities in terms of fitness, gender and learning.

But it is possible to get in the way of accessing wellbeing by distracting from the inherent qualities and affordances of treed landscapes, which emphasises how strong a role environment plays within shaping our motivations for, experience and enjoyment of green physical activity. Active and embodied engagement seems to be an important way to connect with nature, whether running wildly amongst them or simply drinking in the sounds of trees. Moving more can sometimes be a way to encourage more moving experiences in nature.

References

Adcock, C. (1992). Conversational drift: Helen Mayer Harrison and Newton Harrison. *Art Journal, 51*(2), 35–45.

Aronsson, J., Waite, S., & Tighe-Clark, M. (2015). Measuring the impact of outdoor learning on the physical activity of school age children: The use of accelerometry. *Education and Health, 33*(3), 57–62.

Barton, J., Bragg, R., Wood, C., & Pretty, J. (Eds.). (2016). *Green exercise: Linking nature, health and wellbeing*. London: Routledge.

Brymer, E., & Davids, K. (2012). Ecological dynamics as a theoretical framework for development of sustainable behaviours towards the environment. *Environmental Education Research, 19*(1), 45–63.

Brymer, E., & Davids, K. (2014). Experiential learning as a constraint-led process: An ecological dynamics perspective. *Journal of Adventure Education and Outdoor Learning, 14*(2), 103–117.

Cooper, A. R., Page, A. S., Wheeler, B. W., Hillsdon, M., Griew, P., & Jago, R. (2010). Patterns of GPS measured time outdoors after school and objective physical activity in English children: The PEACH project. *International Journal of Behavioral Nutrition and Physical Activity, 7*(1), 1–9.

Friends of Ham Woods. (2014). Available at http://www.hamwoods.org.uk/.

Good from Woods. (2014). *Good from Woods research toolkit*. Available at https://www.plymouth.ac.uk/research/peninsula-research-in-outdoor-learning/good-from-woods/the-toolkit.

Hamer, M., Stamatakis, E., & Steptoe, A. (2009). Dose-response relationship between physical activity and mental health: The Scottish Health Survey. *British Journal of Sports Medicine, 43*, 1111–1114.

Health and Social Care Information Centre. (2015). *Statistics on obesity, physical activity and diet—England*. Available at http://www.hscic.gov.uk/catalogue/PUB16988.

Kaplan, R., & Kaplan, S. (1989). *The experience of nature: A psychological perspective*. New York: Cambridge University Press.

Kellert, S., & Wilson, E. O. (1993). *The biophilia hypothesis*. Washington, DC: Island Press.

Kriemler, S., Meyer, U., Martin, E., van Sluijs, E. M., Andersen, L. B., & Martin, B. W. (2011). Effect of school-based interventions on physical activity and fitness in children and adolescents: A review of reviews and systematic update. *British Journal of Sports Medicine, 45*(11), 923–930.

Markevych, I., Schoierer, J., Hartig, T., Chudnovsky, A., Hystad, P., Dzhambov, A. M., … Fuertes, E. (2017). Exploring pathways linking greenspace to health: Theoretical and methodological guidance. *Environmental Research, 158*, 301–317.

Mitchell, R. (2013). Is physical activity in natural environments better for mental health than physical activity in other environments? *Social Science and Medicine, 91*, 130–134.

National Children's Bureau (NCB). (2013). *Greater expectations: Raising aspirations for our children*. London: NCB.

Ottoson, J., & Grahn, P. (2005). A comparison of leisure time spent in a garden with leisure time spent indoors: On measures of restoration in residents in geriatric care. *Landscape Research, 30*, 23–55.

Pasanen, T. P., Ojala, A., Tyrväinen, L. J., & Korpela, K. M. (2018). Restoration, well-being, and everyday physical activity in indoor, built outdoor and natural outdoor settings. *Journal of Environmental Psychology, 59*, 85–93.

Plymouth City Council. (2014). *Ham Woods*. Available at http://www.plymouth.gov.uk/hamwoodslnr.

Plymouth Fairness Commission. (2013). *An initial presentation of evidence*. Retrieved from http://www.plymouth.gov.uk/plymouth_fairness_commission_introductory_analysis.pdf.

Pretty, J., Peacock, J., Sellens, M., & Griffin, M. (2005). The mental and physical health outcomes of green exercise. *International Journal of Environmental Health Research., 15*, 319–337.

Pryor, A., Townsend, M., Maller, C., & Field, K. (2006). Health and well-being naturally: 'Contact with nature' in health promotion for targeted individuals, communities and populations. *Health Promotion Journal of Australia, 17*, 114–123.

Public Health England (PHE). (2014). *From evidence into action: Opportunities to protect and improve the nation's health*. Retrieved from https://www.gov.uk/government/uploads/system/uploads/attachment_data/file/366852/PHE_Priorities.pdf.

Public Health England (PHE). (2018). *Child and maternal health profile*. Retrieved from https://fingertips.phe.org.uk/profile/child-health-profiles/data#page/1/gid/1938133228/pat/6/par/E12000009/ati/102/are/E06000026.

Quay, J. (2017). From human-nature to cultureplace in education via an exploration of unity and relation in the work of Peirce and Dewey. *Studies in the Philosophy of Education, 36*, 463–476.

Raanaas, R. K., Evensen, H. H., Rich, D., Sjostrom, G., & Patil, G. (2011). Benefits of indoor plants on attention capacity in an office setting. *Journal of Environmental Psychology, 31,* 99–105.

Rogers, S., Waite, S., & Evans, J. (2017). Outdoor pedagogies in support of transition. In S. Waite (Ed.), *Children learning outside the classroom: From birth to eleven.* London: Sage.

Rogerson, M., & Barton, J. (2015). Effects of visual exercise environments on cognitive directed attention, energy expenditure and perceived exertion. *International Journal of Environmental Research and Public Health, 12*(7), 7321–7336.

Sallis, J. F., & Glanz, K. (2006). The role of built environments in physical activity, eating and obesity in childhood. *The Future of Children, 16*(1), 89–108.

Stone, M. R., & Faulkner, G. E. J. (2014). Outdoor play in children: Associations with objectively-measured physical activity, sedentary behavior and weight status. *Preventive Medicine, 65,* 122–127.

Ulrich, R. S. (1981). Natural versus urban scenes: Some psychophysiological effects. *Journal of Environment and Behaviour, 13,* 523–556.

Ulrich, R. S. (1984). View through a window may influence recovery from surgery. *Science, 224*(4647), 420–421.

Waite, S. (2010). Losing our way? Declining outdoor opportunities for learning for children aged between 2 and 11. *Journal of Adventure Education and Outdoor Learning, 10*(2), 111–126.

Waite, S. (2013). Knowing your place in the world: How place and culture support and obstruct educational aims. *Cambridge Journal of Education, 43*(4), 413–433.

Waite, S. (2015). Culture clash and concord: Supporting early learning outdoors in the UK. In H. Prince, K. Henderson, & B. Humberstone (Eds.), *International handbook of outdoor studies.* London: Routledge.

Waite, S., & Davis, B. (2007). The contribution of free play and structured activities in Forest School to learning beyond cognition: An English case. In B. Ravn & N. Kryger (Eds.), *Learning beyond cognition* (pp. 257–274). Copenhagen: The Danish University of Education.

Waite, S., & Goodenough, A. (2018). What is different about Forest School? *Journal of Outdoor and Environmental Education, 21*(1), 25–44. Retrieved from http://link.springer.com/article/10.1007/s42322-017-0005-2.

Waite, S., Goodenough, A., Norris, V., & Puttick, N. (2016). From little acorns: Environmental action as a source of ecological wellbeing. *Pastoral Care in*

Education: An International Journal of Personal, Social and Emotional Development, 34(1), 43–61. Retrieved from http://www.tandfonline.com/doi/full/10.1080/02643944.2015.1119879.

Waite, S., Passy, R., Gilchrist, M., Hunt, A., & Blackwell, I. (2016). *Natural Connections Demonstration Project 2012–2016: Final Report.* Natural England Commissioned report NECR215. Retrieved from http://publications.naturalengland.org.uk/publication/6636651036540928.

Waite, S., Rogers, S., & Evans, J. (2013). Freedom, flow and fairness: Exploring how children develop socially at school through outdoor play. *Journal of Adventure Education and Outdoor Learning, 13*(3), 255–276. https://doi.org/10.1080/14729679.2013.798590.

White, M., Pahl, S., Ashbullby, K., Herbert, S., & Depledge, M. (2013). Feelings of restoration from recent nature visits. *Journal of Environmental Psychology, 35,* 40–51. https://doi.org/10.1016/j.jenvp.2013.04.002.

Wright, N., Waite, S., Graham, L., Aronsson, J., & Waite, R. (2016). *Creating happy and healthy schools through outdoor learning.* Plymouth: University of Plymouth. Retrieved from https://www.plymouth.ac.uk/uploads/production/document/path/10/10803/RFJ27519_Education_folder_and_amends_CORRECTProof_3A.pdf.

Yeh, H. P., Stone, J. A., Churchill, S. M., Wheat, S., Brymer, E., & Davids, K. (2016). Physical, psychological and emotional benefits of green physical activity: An ecological dynamics perspective. *Sports Medicine, 46,* 947–953.

9

Natural Sources of Biophilic Wellbeing

What Is Biophilic Wellbeing?

Biophilic wellbeing can be understood as the ways in which different aspects of human wellbeing are met through passive or active engagement with nature. Several theories and numerous studies, explored in more detail in Chapter 2, contribute to our understanding of biophilic wellbeing, notably biophilia, our innate affinity with the natural world rooted in our shared evolution (Wilson, 1984). The premise is that human fascination with, emotional response towards and general preference for green spaces are evolutionary functions that have long guided our survival and thriving within the natural world. Throughout the Good from Woods case studies, there were examples of people experiencing wellbeing through visual, auditory, olfactory, tactile engagements with the natural world. The pilot framework understood biophilic woodland wellbeing as indicated by feelings of closeness to the natural world, being engaged in a relationship with nature.

A. Goodenough and S. Waite, *Wellbeing from Woodland*,
https://doi.org/10.1007/978-3-030-32629-6_9

Back to Our Roots or Rooted in Nature

As we have discussed in the first chapter on Woodland wellbeing, there are many theories about why humans seem to prefer certain types of natural environment and value the presence of trees in a landscape for shelter and refuge, such as biophilia (Wilson, 1984), environmental preference (Kaplan, 1992), attention restoration (Kaplan & Kaplan, 1989) and psycho-evolutionary theory (Ulrich, 1993). These theories are underpinned by the idea that a natural environment is where humans evolved and our response to such settings today is rooted in the successful reactions of our ancestors to them, ones that allowed them to thrive. Landscapes or even facets of them (such as the colour green) suitable for human thriving may still elicit positive feelings and physiological responses within us, while those posing threat provoke negative or neutral reactions.

Biophilic wellbeing can therefore be explored in terms of how such biophilic reactions and affinities with nature underpin a need to connect with the natural world. Lumber, Richardson, and Sheffield (2017) identify a range of experiences of relatedness to nature that they suggest are indicators of feeling connected with it. These include (ibid., Table 18): (1) pleasurable engagement with nature through the senses; (2) using nature-based metaphor or language to communicate ideas; (3) experiencing a sense of love for nature following nature engagement; (4) feeling compassion and concern towards nature that can support caring attitudes and actions; and (5) perceiving and enjoying beauty in nature. The authors (ibid.) argue that these indicators are also pathways towards connection.

A number of measures to assess our subjective sense of nature connection have been constructed including the short (6 questions) version of the Nature Relatedness Scale (Nisbet & Zelenski, 2013) and the Nature Connection Index (Hunt et al., 2017). Detailed understandings of what meeting our need for nature might feel like emerged after the Good from Woods framework and indicators were first piloted but align with some of practitioner-researchers' findings that are also explored in other chapters. Importantly, they frame connection to nature as achieved not only through direct engagement with the material natural world, but also via more abstract ideas of nature and our place within it. Evidence from within Good from Woods case studies points to how experiences of relation with material nature, its cues to activity and emotion and its immediate and longer-term wellbeing impacts, are negotiated via who we are and what we know about ourselves (memories, knowledge and ideals) and about nature (perceptions of nature's activity and benefits).

A growing set of theories offers alternative or supplementary explanations of our connection with nature. Such perspectives position humans as an inextricable part of nature situated within shifting patterns of relation (Taylor & Giugni, 2012; Thompson, 2016). They suggest preferences for nature may be an entanglement of psychological, physiological, social and cultural responses. Within such approaches, material nature is also repositioned as playing an active role in stimulating and responding to interspecies relations (Gagliano, 2013; Rautio, 2013), and culture and environment may be indivisibly linked in response to natural world settings and sense of our place within them (Waite, 2013). A wide range of theorists have been influential in this reframing of human and non-human relation as part of entangled, fluid communities or 'assemblages' of shared activity (Alldred & Fox, 2017; Taylor & Giugni, 2012). The 'common world' interpretation, for instance, drawing on Haraway amongst others, finds both human and non-human participants mutually transformed within their encounters, the legacy of which is carried into other meetings: links in a 'chain' of 'relationality' (Taylor & Giugni, 2012: 112). Some of these perspectives are discussed in Chapter 1.

In the first case study, we look at how children expressed their relationship with the material and natural world and its support of feelings associated with biophilic wellbeing—how trees became partners in their play.

Case Study 1: Trees as Partners in Play

This research was undertaken in an adventure playground, Fort Apache (FA), situated in woodland regrowth on a former landfill site in Torbay, Devon, in collaboration with Play Torbay. It set out to explore and capture the interactions of children and nature with one another using arts-based methods while they played there and reflect on the impacts on wellbeing. The area is amongst the 10% most deprived neighbourhoods in the country (Indices of deprivation, n.d.).

Naomi Wright, an artist, environmental scientist and playworker interested in nature-based play, who was volunteering as a playworker there, recorded young people's (aged between 3 and 15 years) transactions with the re-emerging woodland. She also noted and described the habits and health of the plant life and its participation in children's play and wellbeing.

The Woodland Site

Fort Apache occupied a thin and sloping parcel of land at the edge of a large housing estate, accessible by formal footpaths and informal gaps in fences and between shrubs. The playground's dominant tree species was sycamore (Acer pseudoplatanus), a vigorous non-native coloniser of vacant land, casting dense shade (GB Non-Native Species Secretariat, n.d.). Historically sometimes perceived as 'weeds' amongst trees, sycamore's resilience, particularly to climate change, is leading to reappraisal of its value (ibid., Binggeli, 1994). Native, field maple (Acer campestre) and ash (Fraxinus excelsior) also grew along its top boundary, with some elder and hawthorn understory. Other flora included native and naturalised species like ivy (Hedera helix), bramble (Rubus fruticosus), alexanders (Smyrnium olusatrum) and old man's beard (Clematis vitalba), alongside escapees from surrounding gardens.

Playworkers (employed by Play Torbay, a voluntary sector organisation) worked at Fort Apache during parts of the week, aiming to support play opportunities at the well-visited site. Volunteering as a playworker, Naomi, the practitioner-researcher, developed her methodology over a year of visits, refining it in response to challenges and success. Her approach drew on her existing expertise and voluntary experiences at Fort Apache, generating evidence from natural and human communities using artistic and playful approaches. In this case study, the innovative methodology was crucial to articulating the interconnections between human and non-human—and how woodland impacted human wellbeing, as humans simultaneously affected the woods, in ongoing unfolding patterns.

Naomi chose to reflect on the relationships between herself, the research methods and the respondents (including the playground) as collaborative, each group involved in shaping a piece of work that could shed light on why young people felt good at FA (Wright, Goodenough, & Waite, 2015). This stance required her to acknowledge her powerful role in ultimately deciding the research methods to be used, but provoked responsiveness and flexibility, requiring her to reshape her approaches in response to the interventions of the other participants.

In practice, this approach included ongoing observations and interpretations of the landscape (terrain, plants and their habits), the children aged between 3 and 15 years[1] (styles of play, types and patterns of movement and rest) and interactions between them, recorded in a sketchbook through written and drawn field-notes. These notes provided a first layer of 'mapping' and inspired a second layer of interpretation, abstract prints that conveyed Naomi's sense of how place, people and culture wove together. Rather than objectively plotted records of action and space, she also noted moods and sensations on these prints. Naomi shared these with children and invited them to record their impressions of Forest Apache, encouraging them to create new maps and add to her prints, making a third layer of evidencing. Some participants also chose to wear accelerometers during a play session and show Naomi exactly where they moved, how often and how intensely, so Naomi could add these records to the other interpretations. A final layer of data was added by Naomi and Alice, the project researcher describing the position, health and habits of the natural features of the site (Fig. 9.1).

Other methods included walking interviews and responsive installations. Walking interviews or 'conversational drift' (Adcock, 1992) were conducted with children as guides, inviting the practitioner-researcher into the spaces and conversations they wanted to share with her (Wright et al., 2015). Through their requirement for the practitioner-researcher and young person to physically navigate the woodland, these child-led tours naturally drew the natural world into conversation, movement and mood, helping reveal their role within young people's enjoyment of Fort

[1]Most evidence collected came from young people aged 10 or 11 years. This may be the most common age when they visit Fort Apache and had freedom to access it independently. It might also indicate that this age group particularly responded to the playful research methods.

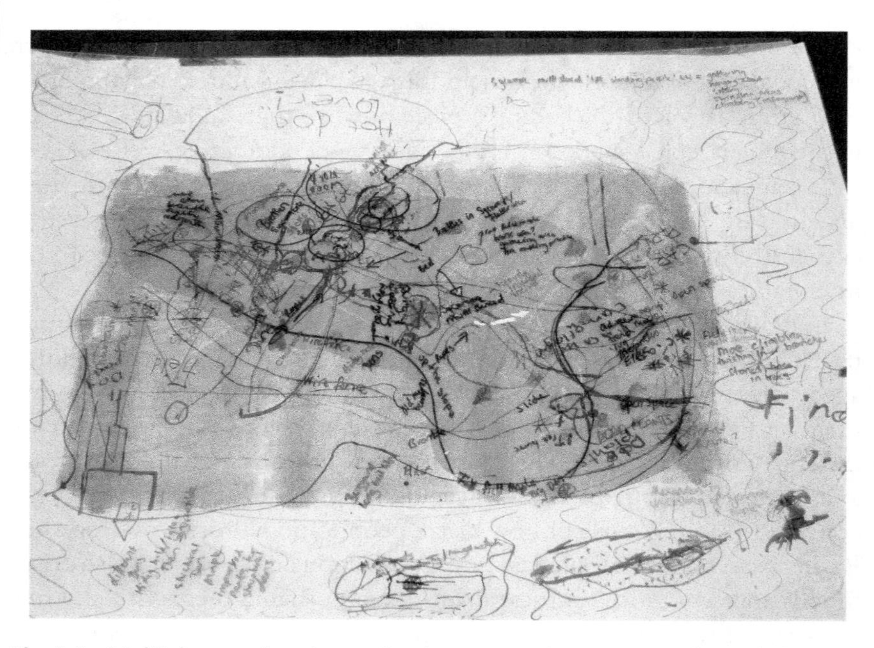

Fig. 9.1 Multiple mapping data collection approach at Fort Apache including layers of information created by the researcher, project researcher and young players (*Source* Fort Apache Local Partner's research evidence)

Apache (ibid). Responsive installations were inspired by children's desires for play and aimed at initiating playful conversations. Seeing the makeshift playhouses children made in trees and shady spaces and equipped with dumped white goods, Naomi furnished a tree house with dressing up clothes and props for a tea party as a statement of her wish to join in play at FA and understand its purposes. Inspired by young people collecting random objects that emerged from the former rubbish tip beneath FA, Naomi fixed an angel she found on a fence post as a stimulus and recorded children's imaginative narratives of its arrival there (Wright et al., 2015).

Trees as Play Partners

Naomi recorded social and psychological impacts of play at FA, but we have chosen to focus here on how players' interactions with nature were supportive of biophilic wellbeing. The diverse natural environment of woodland at FA supported children to experience positive emotional states through both physical and emotional closeness with nature. Young people found many opportunities to play freely and purposefully in this outdoor adventure playground, with and without the support of playworkers. Trees and spaces in and around them were integrated into imaginative games and play scenarios that led to the players feeling good.

One level of engagement with woodland and its environment captured by the practitioner-researcher seems to have taken place at a less conscious level than experiences involving memory, imagination or acknowledged emotion. These interactions with nature appeared more automatic, possibly inspired by a signal from the natural environment processed at a preconscious level (Rautio, 2013). An instance of this was young players' desire to be up in, or amongst, the tops of trees. At FA, Naomi observed, these places provided practical advantages allowing players to witness activity without having to be involved, but also provided opportunities to recline in close alignment with trunks and branches and enjoy relatively green views that were engaging and relaxing (Ulrich, 1993).

> They're just one of my favourites [the hammocks provided by playworkers] ...They make me feel peaceful when I lie down, and you can look up and stare at the sky and all the trees... [Sian, 9].

> [My bed in the branches] on this tree is up this near the fire...it makes me feel good cos it looks like I could go to sleep...I can see a trampoline...cars and vans...people, houses, a fire...trees [Mark, 9].

Mark's 'bed' was in one of two mature, multi-stemmed sycamores, which were used heavily by players for resting and meeting. Space amongst the upright, thick stems allowed young people to lean and find support. The practitioner-researcher and Alice, the project researcher captured evidence of players' heavy interaction with the trees including patches peeled clean of bark, carved initials, paint and pen marks, and scorching from a fire lit nearby (Fig. 9.2).

Fig. 9.2 Scorched sycamore resting place at Fort Apache Adventure Playground (*Source* Fort Apache Local Partner's research evidence)

Naomi found that children commonly snapped twigs and small branches across the site and sticks were moved to other places to furnish play, fire-making and den building. However, sometimes the stick stripping and collecting had no clear purpose but was perhaps a response to the invitation of snappy, woody material and involuntary gratification in breaking and gathering it (Rautio, 2013). Mature sycamore had thick twiggy outgrowths at their base, which were continually snapped and moved away by young people. This created an interactive and circular relationship; as the young people coppiced the basal growth, the tree sent out more twigs to gather more sunlight which provided material for and stimulated more harvesting.

Dens, which the young people constructed or found amongst and beneath trees and shrubby greenery, were sometimes used to intensify contact with nature Naomi's data suggests. Players could hide in dens away from human company and experience a relaxing interaction with nature. Dens could also support restoration through screening out distracting or more stressful contexts, both at Fort Apache and within young people's personal lives.

> That tree was there, and you could sit down on it…if you were a bit upset and you went to Fort Apache you could just go sit in there quietly [Amy, 9].

> There used to be a really good one [den]…that was really hidden but…I like hiding…I like just being where people can't go…like no one uses them, or I can be by myself [Ricky, 13].

> [My den is] kind of a little gardeny area where we can chillax now in a natural area and get away from all the screaming and shouting here and birds tweet which you can hear right now in the background [Carina, 9].

> [I come to FA] To escape from being near my sisters…They wake me up at half one in the morning. I like making dens as well [as playing with friends]…Like in places where you won't see or think of…Up there, that little bush is part of my den as well; so that I can sit in there and have more quietness and if it rains it won't get me wet that much [Amy, 9].

Dens provide quiet, less social places to relax in contact with nature, a refuge and shelter (Ulrich, 1993) and places to restore (Kaplan & Kaplan, 1989). Naomi recorded instances of players protecting their access to these restorative spaces with imagined strategies. One player, for example, said his den could only be entered with a secret code, while another's field maple hideaway, which helped him to 'calm down', was made more secure with stick weaponry and an imaginary guard.

> He kind of looks after my base at night and he does a very good job…I haven't given him one [a name] yet…He's afraid to come out at the morning time, but if he wants to he will [Dean, 10].

A female player at FA taught the researcher how to enter her more inaccessible dens, 'grab hold of nature, grab hold of trees and then you go down with them…they're friendly nature!' [Carina, 9]. Her friendly relationship with nature was also reflected in her daydreaming.

> Daydreaming – [I] love. Sometimes daydreaming that a squirrel might fall into my arm [I daydream most at] Fort Apache…when I'm lying down on my back or sitting down [Carina, 9].

This case study highlighted the role of material nature and imagination in engendering biophilic wellbeing. Interaction with nature was regularly an embodied response to the resilience and possibilities of the wooded site, such that the trees seemed accommodating partners in the children's play, especially when human interactions were challenging (Goodenough, Waite, & Wright, in review). Biophilic effects on children's wellbeing appeared to be gained both passively and actively, players using personal, practical and imaginative strategies for enhancing their interaction with nature.

Top 3 messages from the Fort Apache adventure playground study

1. Children need freedom to create their own play, feel in control and build a relationship with nature, which is accommodating and non-judgemental of their feelings.
2. Less manicured sites permit greater freedom to interact with trees and plants viscerally and directly, which seems especially important for young people that are experiencing challenges in their lives.
3. Interactions are diverse, including snapping or collecting as relaxation through repetition and highly imaginative scenarios stimulated by natural and found objects that enable other ways of being to be envisaged.

While this research captured data about biophilic wellbeing in the moment using mobile and art-based methods, the next case study used more conventional methods of survey and interview, asking young people to recall the experience of tree planting and consider how it made them feel retrospectively.

Case Study 2: Feeling Virtuous About Doing Your Part for the Environment

In this second piece of GfW research, we explore the long-term effects of tree planting on satisfaction of biophilia. The Woodland Trust was keen to establish the medium to long-term wellbeing outcomes of schoolchildren taking part in tree planting, a large area of their work. This UK charity provides tree planting 'packs' and resources to schools with up to 420 saplings of various native species. Trust staff, however, have little direct contact with young tree planters and limited feedback about the outcomes of the experience. For this case study, a retrospective methodology to try and map longitudinal impacts was designed by the Trust, two practitioner-researchers working on their behalf and the project researcher. The aim was to investigate young people's memories and feelings about tree planting in an area where the Trust had organised planting on a fairly large scale in the recent past. Practitioner-researchers were focused on surfacing not only the wellbeing outcomes of tree planting, but whether and how previous planting was associated with their current concepts and behaviours around trees, woods and nature.

The Woodland Site

The location selected for research was in the china clay and St Austell area of mid Cornwall in the far south-west of the UK. Over a quarter of neighbourhoods in the china clay area are amongst the 20% most deprived in the UK, and in St Austell and Mevagissey, one-fifth of neighbourhoods are amongst the 20% most deprived (Cornwall Council, 2015). Woodland Trust tree planting had taken place in a local disused china clay quarry over four consecutive planting seasons, four or more years prior to the research and involved approximately 1500 schoolchildren. Nicky Puttick and Vicky Morris, the practitioner-researchers, identified eight destination secondary schools that these young planters might now attend. The Woodland Trust introduced the project to all eight schools and four agreed to take part.

Nicky and Vicky designed an electronic survey to ask young people to share any memories they had of planting trees and describe their recollection of where, when and why it was and anything else they could remember. The survey was administered by school staff either on paper or using a simple website gateway to an online version for 5–10 minutes during a suitable lesson. Its purpose and staff and students' informed consent for involvement were described in an introduction, and students were asked to confirm their understanding and wish to take part.

The survey results were analysed to indicate key themes to follow up in pupil discussion groups to clarify and enrich the survey data and test researchers' interpretations. 113 young people agreed to complete the survey: 50 male (aged 11–15 years) and 63 female (11–16 years) with two students preferring not to state age and gender. Young people were invited in the survey to sign up for follow-up discussion groups at three schools. Teachers sent information and consent forms home with students wishing to take part in this and eighteen students joined these groups (12 male, 6 female, aged 11–15).

Survey Results

Pupil responses to the survey were mostly brief and factual. 55% of participating pupils could remember planting a tree. Single aspect memories (location for instance) were given by 36 survey respondents, with 23 providing some further detail. Most descriptions focused on the planting location (local school grounds, farmland, the disused china clay quarry, private gardens and youth institution's grounds), or people leading or undertaking planting (primary school staff, family, specialist facilitator, etc.). Other aspects recalled were temporal factors (such as when trees were planted, how long it took), cultural details (how or why to plant trees) and interpersonal elements (who was present with the planter). This type of recall may partially reflect the design of the survey and the short time scale for completion. However, this circumstantial style of recollection would also typify early conversation within discussion groups: 'I planted an acorn once but I haven't been back to it with my parents'; 'Planted a tree with my school at a clay pit', 'You put it in the ground and cover it up again', '…we all worked in pairs and planted a tree', 'We got a free hat'.

Some tree planters reported motivations including improving aesthetics, developing green space, commemorating a pet, person or event and pleasure. Six mentioned feeling good during planting: 'it was so much fun having everyone helping and having a good time, and it is also helping the ecosystem'; '…it's so awesome knowing that you can communicate and have fun instead of work, work, work all the time'.

Memories largely lacked the visual, tactile, olfactory, aural and material detail of tree planting. Nicky and Vicky's analysis found little sensory detail that might be expected to characterise encounters with nature associated with biophilic wellbeing. Further, pupils' initial recall rarely described personal outcomes of planting, such as positive or negative emotions. This does not mean young tree planters did not have engaging, embodied or emotive encounters with the natural world. Responses may have been short and factual because respondents were not sure why they were being asked to provide this information and who or what it was for (Mahon, Glendinning, Clarke, & Craig, 1996; Matthews, Limb, & Taylor, 1998; Nespor, 1998). Without a relevant, personal context for remembering and replaying the tree planting experience, the bare facts were provided.

The practitioner-researchers discovered that like other forms of memory (Michaelian, 2011), recollections of transactions with the natural world were contingent, dependent on the current circumstances of those recalling (as would be demonstrated later in the research when tree planters more fully contextualised memories in the context of their developing scientific knowledge).

An encouraging outcome for the Woodland Trust was that 55% of young people remembered planting a tree. 69% of those memories were recalled with a location and 58% recalled with whom, what or why trees had been planted. Clearly, the act signified a memorable event and through the discussion groups, Nicky and Vicky could probe further for meanings from survey responses that might explain why it had been memorable.

Connection to Nature

The majority of survey respondents selected 'connection with nature' as the most important outcome of young people tree planting, despite few references to the natural world within planters' memories. When asked to interpret connection with nature within follow-up discussion, tree planters described several, sometimes interlinked, ways in which it might take place.

- *Being in the company of animals and the environment*

'Your mind's more open to what's out there and your emotions are much happier when you're with them…Animals'.

- *Taking care of the natural world*

'Sort of like – like you're benefiting to the animals like helping them out to get a long life a lot more'; 'Yeah cos like we use a lot of trees for paper and that, it's like giving something back'.

- *Increasing knowledge and understanding of natural world processes, its requirements to thrive and pro-environmental action*

'Understanding more of how it works and how to protect it'; '...being close to nature is doing anything that helps it and learning more about it – yeah'.

In terms of its impact on mood and satisfaction, 'feeling more calm and relaxed' was selected by survey respondents as the fifth most important outcome of tree planting for young people. However, variation emerged when the results were broken down according to tree planting experience. For respondents who hadn't planted a tree, this was the most frequently anticipated benefit of tree planting (58% of non-tree planting respondents). Contrastingly, only 21% of those who had previously planted a tree chose this option. This possibly highlights a difference between expectations and experiences, and reinforces that memories of immediate, possibly preconscious, impacts to happiness did not retain importance for tree planters. However, the benefit of experiencing yourself in conscious relation to the natural world was an important outcome for both those anticipating impacts and those with experience of them (Fig. 9.3).

Importantly, differences between planters and non-planters were also clear when young people were asked whether it was important to plant trees. 27% of those who had not planted a tree thought it was 'very important', compared with 48% of those with tree planting experience. Taking part in hands on tree planting appeared to incline planters towards feeling the action had meaning and influence. The research team interpreted this result as indicating that providing practical planting experiences to young people can be an intervention towards growing positive dispositions towards tree planting.

On Reflection Through Discussion

Within surveys, tree planting was widely seen as a pro-environmental behaviour. Asked 'Why do you think it is important to plant trees?' 79.8% of respondents selected 'To provide habitats for animals, birds and plants' as one of their three top reasons, significantly higher than any other option. The second most popular choice (44.9% of respondents) was 'to create healthier environments for people'. This finding was confirmed in the

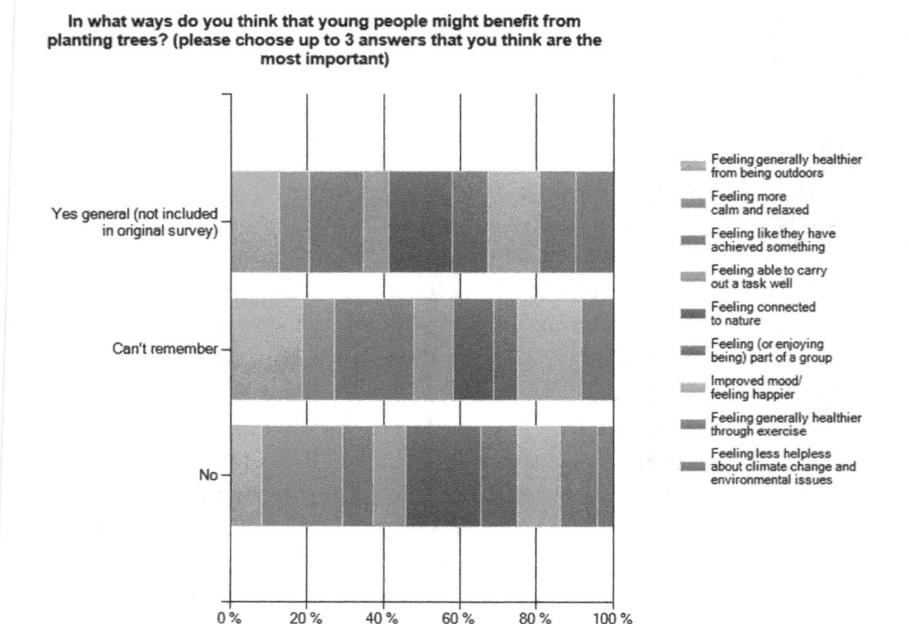

Fig. 9.3 Graph from Woodland Trust GfW case study showing ways in which young people perceived tree planting to be of benefit to their age group. Responses are grouped in relation to tree planting memories: whether students remembered planting a tree in the past (Yes), could not recall if they had planted a tree before (Can't remember) or believed they had never planted a tree (No) (*Source* Woodland Trust Local Partner's research evidence)

discussion groups, and was associated with planters feeling good, including 'a sort of warm glow' and purpose.

> it's like you just slow the world down when I did that [tree planting].

> A good experience to plant that tree…like helping the environment when you're doing it.

> I planted trees with [primary] school when I was eight, and like I don't really feel embarrassed cos I know I was like helping the environment.

> Well, I was sort of just, I just sort of like thought [when planting the tree] it could just be like a future tree later on and just kind of have lots of animals inhabiting it.

Sometimes the memories were revisited in the light of new feelings about tree planting action and its connotations (Michaelian, 2011).

> Not at the time [did tree planting seem significant] …Just I know it was more important now than it was then… it was like two years ago.

> It was on a school trip where we went in a group and just went into the woods it was in primary school and we just planted trees around the area, at the time it didn't feel like it was important, but as I grew up I knew that it was kind of helping the environment and it was important so… because I kind of needed to be more aware of the environment and planting more trees and help the environment and things.

> Well at the time it was in primary and it had no real effect on me because obviously at that age you don't really think about those type of things. But like now in science and you learn about all this stuff and habitats and that you think so you've done like a good deed by helping the environment and that.

Reinterpreting Memories

Developing understandings of environment and society had provided some with newer frameworks for (re)interpreting the meaning of tree planting. Students acknowledged that memories of tree planting were now recalled within a different context and were undergoing 'post hoc rationalisation' (Waite, 2007). Indeed, all tree planting memories are likely to have been reconstructed to more or less extent with increased relevance to students' current experiences and actions (Neisser, 1988; Tarrant, 1996). Revisiting appeared to renew opportunities to experience biophilic wellbeing through tree planting. Engagement in the research seemed to catalyse linking memories and later understandings of the natural world towards an envisioned future (Gough, 1999). Some imagined the future life of their tree (its support of biodiversity, oxygen production, carbon absorption and timber value) and reassess the value of their planting experience from the perspective of the more-than-human world (Gaesser, 2013). These empathetic future-focused narratives, inspired and grounded within tree planting memories, especially developed when pupils considered potential competition between human and more-than-human resource needs.

Many of the good feelings that young people reported regarding tree planting described the possibility of achieving positive emotional moods via altruistic action for the human and more-than-human community. The value of altruism achieved through tree planting was discussed by young people particularly in connection with their ideas about climate change and habitat loss. Planters described their sense that there were currently few opportunities to influence the environmental crises they saw as bequeathed to them by previous generations.

> I reckon it's unfair like cos like when we were like not born like other people were making it worse like, like global warming and we're like our children and that's got to take the like effect of it by other people.

> It's sort of like a war basically. When there's a war, and after the war, the ones that suffer are the kids that are born, and they don't have anything to do with it and yet they suffer.

> '[Tree planting] Makes it like more fair to other people…Because like people are chopping down trees at the moment and like we need trees to breathe our oxygen and they won't have a chance to like live long if we have no oxygen'. 'It's cos they don't care' 'It's like with animals as well'.

The Environment's Wellbeing

Acting in the interests of all of earth's inhabitants and habitats to support our collective future provided positive wellbeing (Emmons, 2003) echoing Zhang, Piff, Iyer, Koleva, and Keltner (2014) finding that time with nature increased pro-environmental attitudes. Respondents in the survey selected 'Feeling like they have achieved something' as the second most likely benefit of young people planting trees. Follow-up discussions revealed tree planters explicitly connected this sense of achievement with self-transcendent motivations and behaviours. 'Feeling able to carry out a task well' was far less frequently chosen, suggesting the achievement of a planted tree was more important than the process. Revisiting the experience of tree planting appeared to help resource a sense of altruistic connectedness with all life forms and attach value to expressions of nurturing them (Fredrickson, 2003). The memory of tree planting and its virtue and value might potentially then provide a resource of biophilic wellbeing when recalled throughout the life course, shifting in relation to developing environmental awareness and experience.

> You're helping the world like have oxygen or something like that... [tree planting could] make you feel kind of successful cos you're making a difference to climate change

> ...Like now in science and you learn about all this stuff and habitats and that you think, you know, you're...so you've like done a good deed [tree planting] by helping the environment and that.

> It feels quite good to feel that like – if you're like making more habitats for animals then like less animals will be extinct...

> It [tree planting] would make you feel you had some part in helping with the whole environment – kind of. You just know that you've done your part and you felt – I don't know – just something.

For some participants, recalling tree planting threatened their sense of wellbeing as they weighed its effects against the scale of current ecological crisis. Exploring tree planting memories in the context of new understandings could instead reinforce a sense of hopelessness.

> Cos it (tree planting) make you feel more guilty by, if you hear about global warming it can make you feel more guilty, having the fact that you've done something but you can't stop like what's happening to global warming.

Overall, however, the survey results demonstrated a statistically significant[2] association between having a memory of tree planting and a belief that tree planting could help address climate change. 27% of survey respondents chose 'to help slow down climate change' as an answer to the question of why it is important to plant trees. Planters were 4 times as likely to select 'to help slow down climate change' (40%) as a reason to plant trees than those who hadn't (10%) or couldn't remember planting a tree (11%). It seems tree planting memories could be recast as a purposeful act in relation to climate change in the context of pupil's increasing knowledge of ecosystem services provided by trees and the ecological crisis. Over time, the meaning and biophilic wellbeing engendered by tree planting and subsequent reflection may have the potential to alter and expand in relation to students' developing understanding of the more-than-human community (Fig. 9.4).

In conclusion, young tree planters experienced feelings supportive of biophilic wellbeing from believing they had nurtured nature through their action. They imagined how their tree may have increased in value to the natural world, providing ecosystem services as it grew. Wellbeing accrued in relation to their growing knowledge of biological systems and threats posed to them by human activity, as they progressed through school. Young people rarely perceived tree planting as directly beneficial for the planter, rather as an action transcending self-interest on behalf of the natural world or for wider human good. Imagination and empathy created scenarios where young people, particularly if they had ever planted a tree, saw benefit from planting as pro-environmental action. However, recalling their environmental action did engender personal subjective wellbeing through satisfying their need to affiliate with other species and providing an opportunity to express concern and compassion for the more-than-human world.

---

[2]Chi-squared test producing a p value of 0.02.

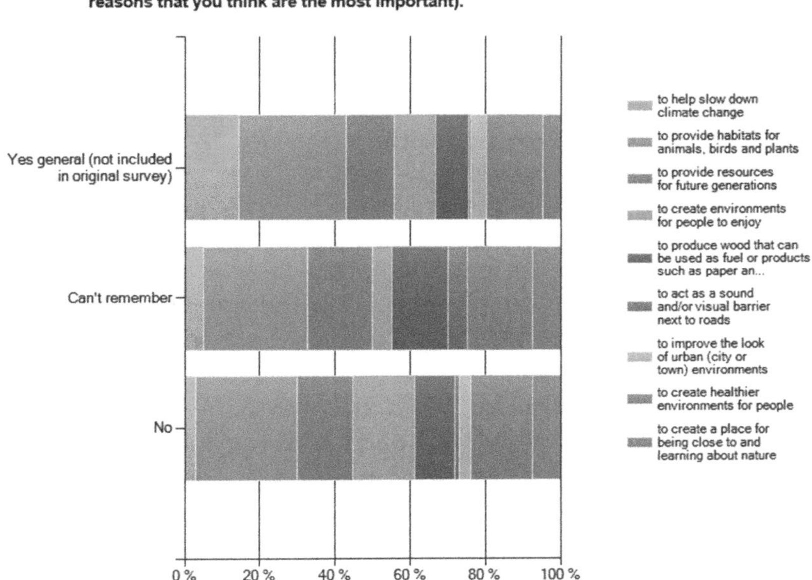

Fig. 9.4 Graph from Woodland Trust GfW case study showing why young people believe it is important to plant trees. Responses are grouped in relation to tree planting memories: whether students remembered planting a tree in the past (Yes), could not recall if they had planted a tree before (Can't remember) or believed they had never planted a tree (No) (*Source* Woodland Trust Local Partner's research evidence)

Top 3 messages from Woodland Trust tree planting study

1. Wellbeing is derived not just from immediate experience but can grow and change in nature over time. Revisiting memories and reflecting on them in the light of new knowledge affect the perceptions and feelings derived from earlier actions, so environmental actions need to be regularly rehearsed and repeated if they are to contribute to biophilic wellbeing effectively.

2. The contemporary reasons for tree planting, such as commemoration, and its process or location seem less important than personal involvement in getting more trees planted and improving the wider environment.
3. Young people are deeply troubled by environmental crises but for most taking even small actions makes them feel more positive about the future and their ability to make changes.

Both case studies drew attention to personal imaginings of self in relation to trees and woodland ecology, posing questions about how much experience of, and empathic relationship with woods we need to establish to reach a point where we can most effectively access and benefit from biophilic wellbeing. They explore the perspective of young people of varying ages and from some of the most deprived neighbourhoods of the UK and some caution should be exercised in generalising these findings to other ages and groups. Nicky and Vicky's research also showed how life stage influenced access to wellbeing with the young people retrospectively reviewing tree planting in the light of growing comprehension of ecology. As described elsewhere, researchers have identified a dip in nature relatedness amongst teenagers. Could it be that nature connection for young people is evolving during this period, perhaps sometimes from a more playful, direct experience (Fort Apache above and Otterhead case, Chapter 8) towards something more serious and abstract? More nuance in the stage and type of nature connection may need to be addressed in designing interventions.

References

Adcock, C. (1992). Conversational drift: Helen Mayer Harrison and Newton Harrison. *Art Journal, 51*(2), 35–45.

Alldred, P., & Fox, N. J. (2017). Young bodies, power and resistance: A new materialist perspective. *Journal of Youth Studies, 20*(9), 1161–1175.

Binggeli, P. (1994). Controlling the invader. *Tree News Autumn*, 14–15. Retrieved from http://www.mikepalmer.co.uk/woodyplantecology/sycamore/britain.htm.

Cornwall Council. (2015). *Indices of Multiple Deprivation 2015 Headline data for Cornwall.* Retrieved from https://www.cornwall.gov.uk/media/15560743/imd-2015-analysis.pdf.

Emmons, R. A. (2003). Personal goals, life meaning, and virtue: Wellsprings of a positive life. In *Flourishing: Positive psychology and the life well-lived* (pp. 105–128). Washington, DC: American Psychological Association.

Fredrickson, B. L. (2003). The value of positive emotions: The emerging science of positive psychology is coming to understand why it's good to feel good. *American Scientist, 91*(4), 330–335.

Gaesser, B. (2013). Constructing memory, imagination, and empathy: A cognitive neuroscience perspective. *Frontiers in Psychology, 3,* 576.

Gagliano, M. (2013). Persons as plants: Ecopsychology and the return to the dream of nature. *Landscapes: The Journal of the International Centre for Landscape and Language, 5*(2). Retrieved from http://ro.ecu.edu.au/landscapes/vol5/iss2/14.

GB Non-Native Species Secretariat. (n.d.). *Sycamore, Acer pseudoplatanus Factsheet.* Retrieved from http://www.nonnativespecies.org/factsheet/downloadFactsheet.cfm?speciesId=25.

Goodenough, A., Waite, S., & Wright, N. (in review). Place as partner: Material and affective intra-play between young people and trees.

Gough, N. (1999). Surpassing our own histories: Autobiographical methods for environmental education research. *Environmental Education Research, 5*(4), 407–418.

Hunt, A., Stewart, D., Richardson, M., Hinds, J., Bragg, R., White, M., & Burt, J. (2017). Monitor of engagement with the natural environment: Developing a method to measure nature connection across the English population (adults and children). *Natural England Commissioned Reports* (Number 233). York.

Indices of deprivation 2015 explorer. (n.d.). Retrieved from http://dclgapps.communities.gov.uk/imd/idmap.html.

Kaplan, S. (1992). Perceptions and landscape: Conceptions and misconceptions. In J. Nasar (Ed.), *Environmental aesthetics: Theory, research, and application.* New York: Cambridge University Press.

Kaplan, S., & Kaplan, R. (1989). *The experience of nature: A psychological perspective.* Cambridge: Cambridge University Press.

Lumber, R., Richardson, M., & Sheffield, D. (2017). Beyond knowing nature: Contact, emotion, compassion, meaning, and beauty are pathways to nature connection. *PLOS One, 12*(5), e0177186.

Mahon, A., Glendinning, C., Clarke, K., & Craig, G. (1996). Researching children: Methods and ethics. *Children and Society, 10*(2), 145–154.

Matthews, H., Limb, M., & Taylor, M. (1998). The geography of children: Some ethical and methodological considerations for project and dissertation work. *Journal of Geography in Higher Education, 22*(3), 311–324.
Michaelian, K. (2011). Generative memory. *Philosophical Psychology, 24*(3), 323–342.
Neisser, U. (1988). Five kinds of self-knowledge. *Philosophical Psychology, 1*(1), 35–59.
Nespor, J. (1998). The meanings of research, kids as subjects and kids as inquirers. *Qualitative Enquiry, 4*(3), 369–388.
Nisbet, E. K., & Zelenski, J. M. (2013). The NR-6: A new brief measure of nature relatedness. *Frontiers in Psychology, 4*, 813.
Rautio, P. (2013). Children who carry stones in their pockets: On autotelic material practices in everyday life. *Children's Geographies, 11*(4), 394–408.
Tarrant, M. A. (1996). Attending to past outdoor recreation experiences: Symptom reporting and changes in affect. *Journal of Leisure Research, 28*(1), 1–17.
Taylor, A., & Giugni, M. (2012). Common worlds: Reconceptualising inclusion in early childhood communities. *Contemporary Issues in Early Childhood, 13*(2), 108–119.
Thompson, J. A. (2016). Intersectionality and water: How social relations intersect with ecological difference. *Gender, Place & Culture, 23*(9), 1286–1301.
Ulrich, R. S. (1993). Biophilia, biophobia and natural landscapes. In S. R. Kellert & E. O. Wilson (Eds.), *The biophilia hypothesis*. Washington, DC: Island Press.
Waite, S. (2007). 'Memories are made of this': Some reflections on outdoor learning and recall. *Education 3–13, 35*(4), 333–347.
Waite, S. (2013). 'Knowing your place in the world': How place and culture support and obstruct educational aims. *Cambridge Journal of Education, 43*(4), 413–434.
Wilson, O. (1984). *Biophilia*. Cambridge, MA: Harvard University Press.
Wright, N., Goodenough, A., & Waite, S. (2015). Gaining insights into young peoples' playful wellbeing in woodland through art-based action research. *Journal of Playwork Practice, 2*(1), 23–43. Retrieved from http://www.ingentaconnect.com/content/tpp/jpp/2015/00000002/00000001/art00003.
Zhang, J. W., Piff, P. K., Iyer, R., Koleva, S., & Keltner, D. (2014). An occasion for unselfing: Beautiful nature leads to prosociality. *Journal of Environmental Psychology, 37*, 61–72.

10

Engineering/Engendering Woodland Wellbeing

As our interdependency with nature is increasingly recognised, policy makers are turning to ways of harnessing its power for societal benefits. As we discussed in Chapter 2, trees are no longer simply seen as timber but as a rich renewable source of ecosystem services and infrastructure for human health. The Research Council UK and DEFRA funded programme Valuing Nature is charting and promoting aspects of these wider benefits of nature (Valuing Nature Network, 2017). The five year, £7m research programme was initiated to improve understanding of the value of nature both in economic and non-economic terms, and to encourage use of such valuations in policy and decision making. Under its goal of human health and wellbeing, projects focus on health improvements associated with urban ecosystems, but can wellbeing from nature be engineered or is it rather engendered? In this chapter, we first define these terms and consider the implications of medicalised perspectives on wellbeing. We then critically examine examples of attempts to engineer human health and wellbeing through nature-based activities, reflecting on barriers and inequalities of access. We conclude that gradual, organic, holistic and sustained nature exposure is necessary to engender benefits, but engineered mediation can ensure appropriate specific interventions for equal opportunities.

A. Goodenough and S. Waite, *Wellbeing from Woodland*,
https://doi.org/10.1007/978-3-030-32629-6_10

To Engineer or Engender

Our title deliberately juxtaposes two forms of 'bringing about' a desired outcome. First, 'engineering' is associated with bringing about a product or outcome through a defined plan that is centrally controlled and generally includes distinct sequential stages. The demand for theories of change in the field of outdoor activities (Fiennes et al., 2015) mirrors this rather mechanistic view of effecting improvements which assumes that influential factors can be manipulated to create a simple model. Engendering, in contrast, evokes a more creative stimulation of desired outcomes, rarely referred to as 'products'. Often engendering gives rise to a broader spectrum of dispositions, attitudes or values rather than a specific and tangible product. Rather than breaking into specific factors, the context is taken to contribute in a holistic, interactive way. These subtle differences in meaning, we argue, have significant implications for how relationships between nature and health and wellbeing are conceived and enacted.

A Cure of All Ills?

Each year, an estimated one in four experience a 'significant' mental health problem in the UK, placing pressure on health service budgets. Prescriptions for anti-depressants and other drugs to alleviate anxiety, for example, are at an all-time high and there are long waiting lists for psychological interventions (Bragg & Atkins, 2016). Equally, there are serious concerns about the nation's physical health with rising levels of obesity, increases in related obesogenic diseases and marked socio-economic inequalities in health (Marmot, 2010). Depression, dementia and social isolation are significant amongst the societal challenges that are placing an increasing burden on the National Health Service and on health and happiness in the UK (Bragg & Atkins, 2016, p. 5) as well as in other parts of the Western world.

In the face of mounting evidence of the benefits of engagement with nature-based therapies and physical activity, government departments have been looking for sustainable and green preventive methods of supporting greater health and wellbeing (Moeller, King, Burr, Gibbs, &

Gomershall, 2018). Public Health England has endorsed these moves to increase use of preventive nature-based health and wellbeing interventions (PHE, 2017). However, limited access to natural environments for sections of the UK population is a key barrier to these aspirations.

According to a Parliamentary briefing note (POSTnote 0538, 2016), DEFRA has estimated that if everyone had access to enough green space, resultant increased physical activity levels could save the health service £2.1 billion. In support of such estimates, a UK cross-sectional study (Mitchell & Popham, 2007) found people who live within 500 metres of accessible green space are 24% more likely to achieve 30 minutes of physical activity a day. Yet 82% of the UK's population now live in urban environments (POSTnote 0538, 2016). Importantly, inequalities in health are exacerbated by this unequal access to green spaces. The need for some form of action is clear when recent research suggests that in the most deprived groups, mortality rates are halved in areas with the greenest space (Marmot, 2010). However, mere availability is insufficient as it is subjective access rather than objective proximity which governs activity levels and use of green space (Hillsdon, Panter, Foster, & Jones, 2006; MENE, 2018). Supporting uptake of access to nature could therefore make a huge contribution to alleviating health inequalities.

Pretty, Rogerson, and Barton (2017) calculated the annual costs of health problems arising from modern lifestyles as £183.6 billion, with mental health costs around £100 billion, so there are clear economic drivers for the adoption of supported access as a cheaper way of providing preventive or rehabilitation treatments. Therefore, a holy grail for research has been to determine minimum dosages for beneficial effects to further drive down the cost of intervention. 'Ecominds', for example, a programme of nature-based interventions, was calculated as saving over £7000 per participant through reduced costs to the National Health Service, reductions in benefit claims and increased tax contributions through returns to work (Vardakoulias, 2013).

However, medicalisation of green health interventions positions wellbeing as 'lack of illness', while according to Jorgensen and Gobster (2010), it is a positive and far broader sense of feeling good in body, mind and soul with associated community and social outcomes. Conceiving of natural environment as green public health infrastructure aligns better with this breadth and the definition of wellbeing used by DEFRA: 'a positive physical, social and mental state; it is not just the absence of pain, discomfort and incapacity. It requires that basic needs are met, that individuals have a sense of purpose, and that they feel able to achieve important personal goals and participate in society. It is enhanced by conditions that include supportive personal relationships, strong and inclusive communities, good health, financial and personal security, rewarding employment, and a healthy and attractive environment' (DEFRA, 2007, cited by Bragg, Wood, Barton, & Pretty, 2014, p. 9).

We noted in Chapter 2 that interventions have been designed to increase woodland in urban and disadvantaged areas to provide infrastructure to meet this public health need, but as we have argued previously, social, cultural and economic mediators amongst different sections of the population impact on perceptions of green space and ultimately access to potential biophilic, emotional, social, physical and psychological health outcomes. This influence of multiple factors makes it difficult to find easy one-size-fits-all solutions and underpins key differences in the services provided by green spaces and green care. While urban greening of parks and streets creates green infrastructure for public health, 'Green care: nature-based therapy or treatment interventions [are] specifically designed, structured and facilitated for individuals with a defined need' (Bragg & Atkins, 2016, p. 18).

Creating Green Infrastructure

The Forestry Commission Scotland's Woods In and Around Town (WIAT) programme in deprived urban communities aimed to regenerate and promote local woods as secure and accessible places in order to increase community contact with woodlands. Woodland engagement by the local community in WIAT sites resulted in significant changes in perceptions of their safety, their use, and attitudes towards woodlands as places for physical activity while no significant changes were observed in the non-intervention control site (Ward-Thompson, Roe, & Aspinall, 2013). Changed access patterns were reported alongside some improvements in wellbeing (Silveirinha de Oliveira et al., 2013), although contrary to expectations stress levels increased in the intervention community (Ward-Thompson et al., 2019). There was a small increase in how often people visited natural environments and those who did so were likely to be less stressed than those who did not. The authors concluded that external factors may have operated and that a larger and longer study would be needed to explore potential benefits (ibid.). Ten years earlier, Kuo (2003) demonstrated the impact of enhanced green space on societal challenges in the series of studies in Chicago explored in Chapter 2, linking tree and grass cover with greater use of residential outdoor spaces by adults and children. Kuo (2003) attributed this to these shared areas becoming 'defensible space', a residential setting whose material character supports residents in safeguarding its security. Accompanying this greater sense of safety, healthier patterns of children's outdoor activity, more social interaction amongst adults, healthier patterns of adult-child interaction and supervision, stronger social ties and greater resource sharing amongst adult residents were all noted. Such changes affirm how culture and place together create structural circumstances for the promotion of wellbeing. Returning to the health savings estimated through greater nature exposure (over £180 billion [Pretty et al., 2017]), the estimated £2.7 billion annual cost to maintain all UK green space (POSTnote 0538, 2016) looks like a small price to pay.

We also now have better estimates for how much time people should aim to spend in nature for benefits to be felt; White et al. (2019) demonstrated that individuals spending two hours or more in nature in the previous week were significantly more likely to report good health or high wellbeing than those who had spent less time. Positive associations peaked between 200–300 minutes per week with no further gain above that. Although the relationship is associative not causal, the fact that the figures also applied to people with long-term illnesses/disability suggests that it was not simply due to healthier people visiting nature more often, but demographic variations were noted, reinforcing awareness that personal motivations, cultural practices and place meanings also play significant contributory roles (Lachowycz & Jones, 2013; White et al., 2019). Different dosage was found by Barton and Pretty (2010) in their metanalysis of ten studies, where shorter exposures (5 minutes of green exercise) were associated with greater gains in self-esteem and mood. Frumkin et al. (2017, p. 6) comment 'commonly used exposure measures cannot quantify the "dose", that is to say, what a person experiences during an episode of nature contact. If two people—one observant and highly attuned to nature, the other oblivious or distracted—both walk down the same forest path, they are likely to "absorb" differing levels of nature'.

Levels of biodiversity in green space and degree and dimensions of greenness as previously explored are also suggested factors in wellbeing outcomes (Brindley, Jorgensen, & Maheswaran, 2018). Lin et al. (2017) note, for example, that gardens with tree cover encourage people to spend more time in natural space with potential health benefits. Astell-Burt and Feng (2019) find that exposure to tree canopy coverage of 30% or more is associated with lower psychological distress and better reported overall health in a longitudinal Australian case study. The authors recommend that 'Protection and restoration of urban tree canopy specifically, rather than any urban greening, may be a good option for promotion of community mental health' (ibid.).

Another dimension worthy of consideration is at what point in the lifespan interventions have the most positive and lasting effect (Hughes, Rogerson, Barton, & Bragg, 2019; Wells & Lekies, 2006). Early outdoor experience has been associated with subsequent positive attitudes to nature and the environment and continuing use of outdoor contexts to promote personal wellbeing (Davis, Rea, & Waite, 2006; Morgan & Waite, 2017). Tillmann, Tobin, Avison, and Gilliland's (2018) review about nature and mental health in children and teenagers (0–18 years), categorised quantitative research papers into three types of nature interaction: accessibility, exposure and engagement. They concluded that although there was evidence of a beneficial influence of nature on children's and teenagers' mental health, a substantial minority of studies found no significant effect. Hughes et al. (2019) discerned a dip in nature connectedness around age 15–16 years as described previously, noting this decline coincides with increasing exam pressure in the UK and that social, cultural and life-stage may influence the visibility of connection to nature.

Lachowycz and Jones (2013, p. 64) suggest a socio-ecological framework for understanding wellbeing impacts of green space, taking into account exposure, demographic, cultural and environmental factors, alongside opportunity, ease and motivation to use, as factors moderating different attitudes towards and uses of green space and its psychological and physical health benefits. In their theoretical framework, causal pathways are unidirectional, but they acknowledge that the relationships between these elements are actually more complex and interactional. For example, socio-economic group affects likelihood of accessing nature; use of green space may affect perceptions of the local environment; both of which in turn may influence uptake of nature-based interventions. They argue that multilevel interventions, where environmental improvements are accompanied by group-specific targeted programmes that support access appropriately, are most successful in changing behaviour and improving health.

A multilevel approach also helps to overcome a tension with medicalisation of health and wellbeing which tends to locate difficulties within the individual. As we have argued earlier in this book, poor wellbeing and health are not just an individual's problem but reflect wider cultural and societal structures that constrain personal welfare (Walton, 2017).

Green Care Interventions

Researchers have also collated evidence to encourage health practitioners to take up nature-based prescribing (Bloomfield, 2017; Bragg & Atkins, 2016; White et al., 2019), but research methodology often does not comply with medical clinical trial models and has thus far failed to persuade large numbers of health practitioners to adopt the practice. Moeller et al.'s (2018) international review of institutional (e.g. hospitals) nature-based interventions found three main types of therapy: horticultural, animal-assisted and care farms; while Bragg and Atkins (2016) identified the most commonly occurring forms of green care in the UK as social and therapeutic horticulture, environmental conservation interventions and care farming.

Bloomfield (2017) used case studies of eight nature-based green prescription interventions to shed light on the necessary dose, replicability (at scale and across contexts), access and cost-effectiveness to address mental health inequalities. His 'Dose of Nature' project was a partnership between health staff, natural environment organisations and intervention practitioners. Each programme ran for 12 weeks for 2–3 hours per week in diverse natural environments with between 4 and 10 participants. Activities varied from active conservation to quiet contemplation. The diversity in place and activities made it impossible to attribute change to specific environmental or mediation factors. Nevertheless, there was an average increase of 69% in self-reported wellbeing in the participants, and two new self-organised support groups were established following the intervention.

Within a comparable collaborative public health project involving two National Parks (Dartmoor and Exmoor) in South West England, the Moor Health & Wellbeing evaluation (Howes, Edwards-Jones, & Waite, 2018a) aimed to identify processes and factors which inhibit or support the successful use of nature engagement across the National Parks to promote health and wellbeing. In addition to confirming positive wellbeing outcomes for participants, the findings indicated that successful outcomes are dependent upon:

- Developing relationships with community groups, health and environmental organisations
- Excellent clear communication between all partners and participants
- Joint ownership & community engagement
- Link workers directing referrals to appropriate services
- Link workers brokering access to natural environments, social opportunities and novel experiences.

There are some studies that map the benefits of certain activities in green space against the needs of particular health service users. Zhang et al. (2017) examined effects of nature-based experiences for people with reduced mobility, finding a preponderance of studies about the elderly. They found walking and gardening were the most common activities. Benefits seemed to derive from active participation in various activities. For example, walking in nature settings had more positive impact on physical symptoms, than walking elsewhere, for people with Parkinson's disease. They also found more vegetation in outdoor green spaces such as trees, alongside wildlife, sunshine and fresh air, were valued influences upon physical, mental and social health benefits for those with limited mobility.

A similar enjoyment of fresh air was observed in a specifically woodland wellbeing project, which was a partnership scheme between Bristol Dementia Wellbeing Service (DWS) and Forest of Avon Trust (FoAT) to enhance the wellbeing of people living with dementia and their families/carers. The programme involved nature-inspired crafts, foraging, photographing their natural surroundings, writing and sharing poetry and chatting together outside around the fire. Both their carers and the people living with dementia found the experience sociable and relaxing, a welcome change from their usual routine (Gibson, Ramsden, Tomlinson, & Jones, 2017). Zhang et al. (2017) argue that the mechanism of achieving benefits from green space use is holistic and multisensory and linked to an empowering sense that other ways of living are possible.

Many of these interventions are delivered through partnerships between the health sector and third-sector service delivery organisations (like many of those whose activity is the subject of their GfW case study research). As Bloomfield (2017, p. 83) comments there are '(at least) two different "languages": the language of healthcare and practice; and the language of nature and environmental engagement'. Bragg and Atkins (2016) make several recommendations for successful partnerships including: clear distinction between general public health programmes and targeted green care; agreement on what green care includes; collaboration within the green care sector to create a coherent and well communicated offer to commissioners, with robustly evidenced examples; and a register to allow easy choices and access to services.

Howes, Edwards-Jones, and Waite (2018b, p. 10) summarise in a nutshell similar steps for a successful green intervention (developed from the Moor Health and Wellbeing project described above). Barriers to referrals from Health Care partnerships included: fears about health and safety; worries about the sustainability of the service; a lack of familiarity with such services and competing priorities for time (Howes et al., 2018a). A vital factor was the support offered by link workers who could talk to potential participants sensitively, help environmental sector partners to plan appropriate activities, and had credibility and authority within the health sector.

Mutual Benefits

Many environmental organisations such as the Forestry Commission, TCV, Groundwork, The Wildlife Trusts, the RSPB and the National Trust have become acutely conscious of growing concerns about societal challenges and there is a discernible shift in their policy and practice from being *for* nature (see Husk, Lovell, Cooper, & Garside, 2013 for a systematic review) to being *with* nature, including developing people's nature relatedness, or nature connectedness and their personal and social wellbeing (Lumber, Richardson, & Sheffield, 2017). The assumption is that by developing an attachment to nature, not only will human wellbeing be served but pro-environmental attitudes will also grow. Direct experiences of nature and engagement with natural beauty catalyses environmental knowledge into pro-environmental motivation and helps overcome the gap between belief and action (Bragg & Atkins, 2016; Richardson & McEwan, 2018).

For Bragg et al. (2014, p. 21), one step towards wellbeing is: 'Being able to *Give*, through sharing and supporting each other and working as a team, by volunteering their time and also by giving back to nature through shaping and restoring natural environments'. Typhina (2017) notes that a range of positive and negative emotions influence pro-environmental behaviour but a lack of affect for or time in nature leads to indifference. Her interviews with urban park visitors explored how and why individuals communicate about and interact with nature and found that social media and sharing experiences could amplify positive effects. Visitors shared significant and memorable sensory experience from use of hammocks under trees, water features, and native plant and animal habitats. Taking this approach rather than just focusing on conservation volunteering capitalises on self-interest and so supports sustained action. As Bloomfield (2017, p. 84) argues: 'It is important to identify mutual benefits, questions that both sides of the equation are interested in answering, and practical solutions that meet everyone's needs'.

The plethora of organisations now involved in attempting to deliver these aspirations can, however, result in communication difficulties with different terms for activity and benefits, different delivery models and different impact assessment measures. Organisations have commissioned researchers to help define contributions more clearly. For example, the Wildlife Trusts asked the University of Essex to carry out a literature review of wellbeing benefits from natural environments rich in wildlife in 2015, which confirmed the evidenced link between high quality biodiverse natural environments and positive health and wellbeing. This was followed in 2016 with an assessment of the contribution to people's wellbeing across 17 Wildlife Trusts' activities for both the general public and individuals with defined needs. They found participants in general developed skills and felt healthier with raised physical activity levels; while people experiencing social disadvantage or poor mental health and unemployed individuals reported improved confidence, self-esteem and mood. The third phase of the research in 2017 looked specifically at volunteering with the Wildlife Trusts, using a health and wellbeing evaluation tool to collate evidence. It found that 95% of people with poor mental health improved significantly after 12 weeks of volunteering with Wildlife Trusts, particularly those who had not previously taken. At the start of the evaluation, 39% of participants reported low wellbeing, compared to UK norms, which reduced to 19% after 12 weeks. Volunteers also said they felt more positive, healthy, connected to nature and their levels of pro-environmental behaviour, physical activity and contact with green space had all increased. Through this drilling down, the environmental sector is becoming more skilled at designing interventions that will meet the demand for mutual benefits.

Designs may be based on an engineering model or seek to engender change. In the following case studies from Good from Woods, we explore the outcomes of engineering and engendering wellbeing from natural health services.

Case Study 1: Engineering Confidence for Family Forest Visits

This case study concerns a three-year project by the National Trust and Family Learning (FL) to encourage families to visit National Trust properties in Devon, Cornwall and Torbay, which was funded by the Big Lottery.

Around 33% of the project's outdoor time was spent in a woodland environment. The intervention sought to reach parents and children referred by Children's Centres from areas of social disadvantage or with learning, social or emotional differences.

The Woodland Site

They visited Halwell Woods near South Pool, Overbeck's woodland area near Salcombe and Gallants Bowers near Dartmouth for bug hunting, mirror walking, smelly cocktails games and stick men making.

Halwell Woods is a dense, ancient semi-natural woodland, with saplings up to veteran trees present and is 1.4 miles from the nearest small settlement. Ground flora is varied but includes many bluebells and ramsons. Overbeck's Woodland is similar in profile and 2.2 miles from Salcombe. Gallants Bower is a more recent wood on very steep ground, with no veteran trees. It is densely treed with good ground cover and 1.8 miles from Dartmouth.

Each session included a woodland walk, a fire and refreshments. Families could return to the sites in their own time and were encouraged to use the skills they had acquired to benefit their local community.

Thirty-two adult learners with their children were involved in the Good from Woods research. Jade Bartlett was the practitioner-researcher, who worked with the National Trust at that time. The focus of the research was to explore what types of wellbeing were fostered through visiting National Trust woodland sites and engaging in staff-led activities (Goodenough, Waite, & Bartlett, 2015).

Most stakeholders (National Trust and Children Centre staff) expected participants would have a positive emotional experience taking part in woodland activities. Supporting parents with their children in the sometimes-unknown context of woodland was expected to build parental sense of competency.

Jade used short mobile 5-minute individual interviews during the woodland activities to collect data during these one-off sessions. This meant that parents could still engage with activities and look after their children, but meant it was sometimes challenging to establish a good researcher-respondent relationship. Some parents seemed shy and pursuing lines of interest was challenging in such short dialogues. However, common themes emerged across the interviews within the framework of wellbeing indicators described in Chapter 3. Jade noted that parents and carers tended to answer questions primarily with their child's wellbeing in mind, rather than their own. This type of response however provided a clue to how adult wellbeing was largely achieved—through feeling good about their children and role as parents. Perceptions of benefits for their child led to an indirect wellbeing benefit for the parent in terms of their competency, confidence and control. Making a good choice on behalf of their child through supporting their access to nature was a source of psychological wellbeing, supporting parents to feel competent. Parents also appeared to feel they had developed by learning new activities to do with their children, learning about new places to go with their children, observing their children learning new things and enhancing their family life by learning about each other and each other's abilities:

> I mean to be honest with you, I'd have never thought of coming here with a picnic so that's something, that's nice so I'll do again. Yeah and the mud-slinging – I'd never do that – too messy! Lazy mummy… [Lucy, Mum with son].

Feeling good was regularly associated with shared parent–child activity: 'It's just the chance to get out, spend some time with [child's name], just the two of us, for bonding, that kind of stuff really' and was also linked with intention to repeat the experience:

> I think it's great for the children, it's wonderful for them to be in nature in the outdoors, I think it feeds their soul, it feeds their imagination, it's great for them to learn about things so they can get an appreciation of it and they will come to love the environment and look after it as they get older [Robin, Mum with daughter].

Respondents also felt good about having opportunities to experience pleasure, aesthetic enjoyment and playful interaction with the outdoors, the fresh air and the woods, 'Because nature is a natural playground'. Fewer attributed satisfaction to the activities that took place within nature.

However, participants did describe a feeling of wellbeing through being supported to visit nature in a group with other parents and perceived wellbeing benefits for their children from their interaction with other children and adults. As Lucy expresses above, they also valued staff leadership and group dynamics in managing children's engagement with the outdoors:

> when I'm on my own with him or something...I'm always thinking right, got to do more things with him all the time, but when you're with a group, it's somehow a bit more relaxing, as long you can just go with the pace and they're telling you where to go next and what to do next, which I think is better [Dave, Dad with son].

> It was tough with a little two year old, I wouldn't have thought they could have done it, but with the groups doing it, they've done it...They [children] take the discipline better from the leader especially when they've started to go to school and it's good for them [Joan, Grandma with grandson, granddaughter and child-minding one girl].

Supported engagement also gave parents a chance to closely observe and witness their child in nature-based experience. It appeared seeing their child feeling good through restorative connection with the natural world was a source of them feeling good.

> We would just walk through them [woods], you know, have a circular walk rather than stop and take it all in, whereas if you do something like this and spend some actual time here you get much more out of it...we wouldn't have done stuff like this [Kirstin, Mum, with daughter].

The intervention was engineered to promote the use of woodland sites for family cohesion and wellbeing, but its effects were only measured in the moment and it is impossible to say what effects might endure in the future.

Top 3 Messages from the Families in the Forest

1. Supporting opportunities for family access to woodland sites and activities allows parents to feel confident that they are enabling their children's growth and development.
2. Seeing their children engaging in a relationship with nature and having the opportunity to 'get out' with their families helps parents to feel good.
3. Being with other families enables parents to feel confident, accepted and safe. They feel supported with expert staff on hand. Parents observe social wellbeing benefits for their children through interactions with other children and adults, and in sharing experiences with their children, parents feel more confident and competent within their own families.

Theory of Change—Engineering

The Moor Health and Wellbeing project (Howes et al., 2018a, p. 67) also used a theory of change related to an engineering approach, where there were targeted groups and tailored activities that enabled the development of green care. An engineering model assumes that steps are relatively discrete and linear to map a logic model for change (Fig. 10.1).

Setting out inputs and outputs offers clarity about intention and enables process and assumptions to be examined transparently but tends to simplify what can be highly complex and volatile contexts.

Case Study 2: Engendering Sustainable Outcomes at Folly Wood

Stroud Woodland Co-op was established as a way for a group of local people to own woodland for the benefit of the environment, shareholders in the co-op, and other local people. The co-op bought one woodland, Folly Wood, at auction in 2010, with each of the 64 members investing

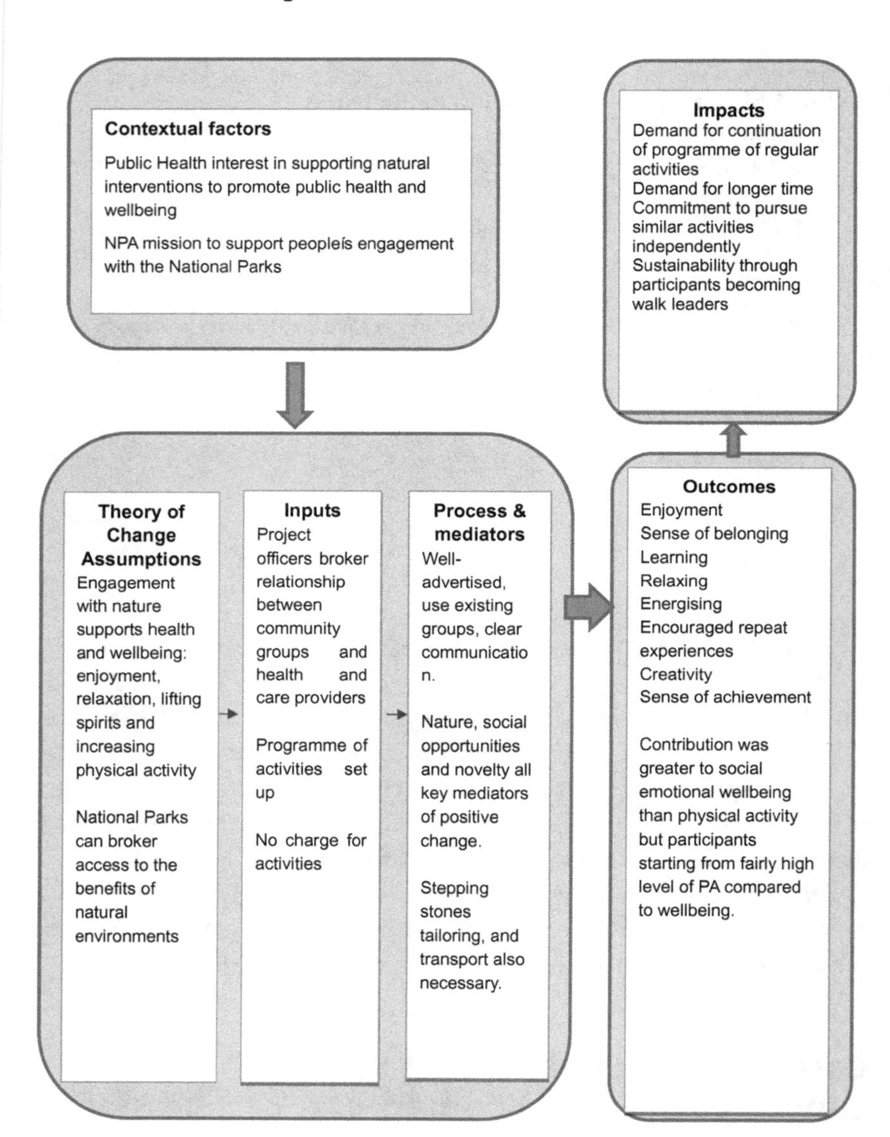

Fig. 10.1 An 'engineering' green care theory of change (*Source* Moor Health and Wellbeing project, Howes et al., 2018a, p. 67)

£500. Shares are owned by individuals, families and organisations. Directors of the organisation are appointed annually and meetings (open to the whole membership) were held to develop plans for the wood, including seasonal group gatherings and celebrations. Members use the woodland for recreation, allow use by local projects/organisations and manage it for biodiversity.

Research at Folly Wood was undertaken by four shareholders in the co-operative, Ali Coles, Seb Buckton, Richard Keating and Jackie Rowanly. They took on the practitioner-researcher role to explore wellbeing outcomes of Folly Wood related to co-operative ownership.

The Woodland Site

The three-and-a-half-acre woodland is at the edge of Stroud in the Cotswolds Area of Outstanding Natural Beauty. Probably remnant ancient beech woodland, Folly Wood had been densely planted with larch over 90 years ago and was unmanaged. The ground flora was predominately ivy, bramble and sweet woodruff. There were active badger setts on site. Clearance work, replanting and provision of simple play and other facilities began in 2012.

The intention was that the community ownership model would provide a range of wellbeing benefits. These were anticipated to be connected to the access to woodland that Folly Wood would provide shareholders, the impact of group membership and collective processes of managing it.

Data was gathered in a range of ways. An online survey asking co-op members and ex-members to rate reasons for becoming involved in the co-op, with questions on demographics, frequency of visits and activities taken part in. A follow-up discussion group of seven members and interviews with three shareholders (one a couple) explored findings emerging from the survey. A creative research workshop including role-play and art activities also aimed to gather impressions of Folly Wood experience. The Good from Woods framework and indicators were used for analysis and reporting (Fig. 10.2).

Fig. 10.2 'Free to go.' Young person's drawing from creative research workshop art activities at Folly Wood (*Source* Stroud Woodland Coop Local Partner's research evidence)

Good Feelings from Folly Wood Now and for the Future

The most commonly recorded indicator of woodland wellbeing was positive emotions and moods (117). Woodland management, the creation of woodland structures and intergenerational activities involving adults and children contributed to feeling good, but so did the processes, practices and sense of joint investment through collective ownership.

> I've always wanted a piece of woodland for myself, but…on my own wouldn't know what to do with it…so there was that "oh, what a great idea, I can join in". I haven't managed to do as much as I'd like to for various reasons, … hopefully if I get to do some work in the woods, that will then carry on even if I'm not around anymore, that's a good feeling, because we never really get to own anything, we just borrow it for a while, don't we, while we are here (FW1).

Feeling purposeful were noted 47 times in the data, sometimes as individual purpose, but commonly as shared purpose, such as the requirements of collectively taking care of the woods for the future.

> It's the sense of purpose that's really kind of tangible and concrete…it's very simple, so the wood is owned by all these people and one of the main purposes is actually to come up and maintain the wood.

Connecting with others through shared beliefs and outlook was equally prevalent, for example by identification with aims of 'communal activity and environmental concern', suggesting entangled mutuality for Folly Wood's arboreal inhabitants and humans.

> …the fact that there should be a woodland available for people, and for it to be a shared facility and [the wood] protected, that's the important thing (FW2).

> …there is a community here, so we have people as well as trees (FW3).

> [I joined] to own some land with some like-minded people and therefore to work with nature in a communal way (FW4).

'Contributing to the greater good' was present at similar levels and echoes the wellbeing derived from tree planting at a later point in their lives in the Woodland Trust case study in Chapter 9. The timescale for this wellbeing is conceived as much longer than the immediate future. Here, 'giving others the opportunity to enjoy woodlands' and the opportunity to 'invest in the community and landscape' suggested wider and longer benefits than the just for the individual (Fig. 10.3). As in the Woodland Trust case study, this indicator frequently related to an imagined future for the trees of Folly Wood:

> It's exciting to think we can leave it for loads of generations and take that sense of stewardship to create something for future generations (FW5).
>
> we don't really know how it's going to –what will happen, that's quite exciting, isn't it? Now the woodland will evolve –it's a living and changing thing (FW9).

Biophilic wellbeing was exemplified through pleasant and appreciative feelings towards other life forms and particularly natural beauty: 'how precious they [badgers] are'; 'it's just lovely [light on leaves], it's just beautiful… the natural patterns and puddles of light and sort of more organic shapes'; 'I love to hear the tapestry of different sounds'. Members described forms of natural connection:

> when you go into the wood it's a different kind of being at one with nature I think, which is really a pleasure to enjoy (F6).
>
> I actually connect on a more pagan level, on a nature level, all those other things fall away and I'm there, it's a much deeper connection, I can understand how the Druids might have worshipped trees, you know I feel it, I don't have to think it (F8).

Fig. 10.3 'New life, new growth, fresh start…' Grandmother and grandson's drawing from creative research workshop art activities at Folly Wood (*Source* Stroud Woodland Coop Local Partner's research evidence)

Top 3 Messages from Folly Wood
Community ownership of woodland delivers benefits through:

1. Knowing that you are contributing to preserving and enhancing woodland and increasing wellbeing benefits for others and 'the greater good' even if you never visit the wood yourself.
2. Feeling safe and free to visit, enjoy and take action based on having had a stake in agreeing what is permissible and desirable with the rest of the group.
3. Engendering social, political and environmental connectedness.

Though less frequent, experiences supportive of social wellbeing (often giving and receiving social support) and physical wellbeing (particularly enjoyment of physical activity) were also mentioned by some.

A Note of Caution

Participants in a community-owned woodland may already possess a strong sense of nature relatedness, motivating them to invest in nature. However, ownership of Folly Wood appeared to allow members to find diverse pathways to pursuing this connection over time. For some, this might simply be contributing financially to support others' conservation and use of woodland, while others spent much more time in the woods.

The cost of a share in Folly Wood would have made co-operative ownership out of reach for many and suggests that membership was largely affluent. However, the principle of a sense of collective ownership and engagement with the woodland appeared to support diverse types of particpation as described and result in a strong outcome of sustainable wellbeing—one that mutually nurtured human and non-human alike. This model of intergenerational community engagement with a woodland is one way that wellbeing can be engendered and has also been successfully implemented in deprived communities (Richardson, Goss, Pratt, Sharman, & Tighe, 2013).

Theory of Change—Engendering

While a logic model is suited to tightly defined engineered interventions, a theory of change related to engendering seeks to acknowledge that there is a coalescence of influences and that outcomes may differ for different groups and at different times. In the Woodlands In and Around Town (WIAT) evaluation of an intervention to create a green infrastructure that could be engaged in diverse ways by individuals, Ward-Thompson et al. (2019, p. 5) demonstrate that initiatives to make sustainable change require an **eco**logical rather than a logic model to acknowledge complexity (Fig. 10.4).

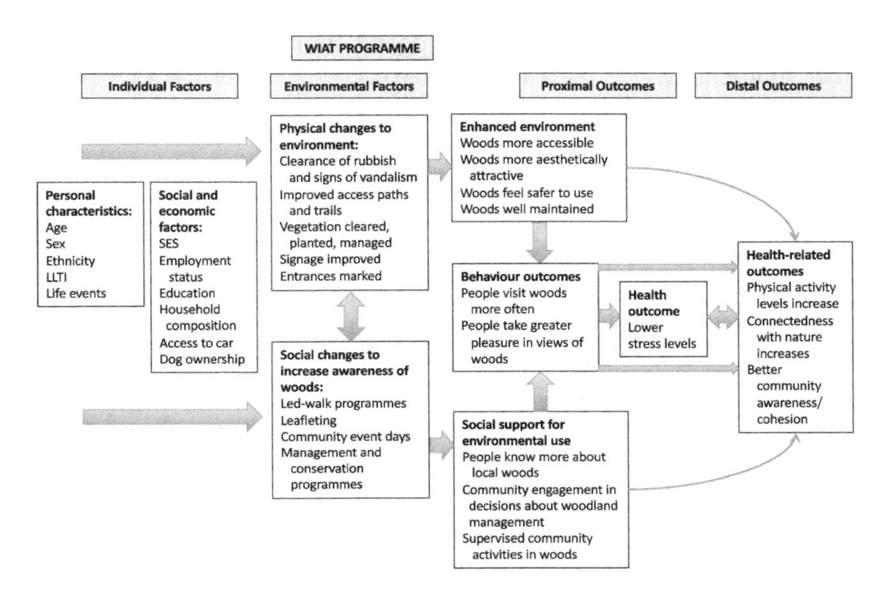

Fig. 10.4 An 'engendering' green care theory of change (*Source* Figure 1: Hypothesised impact pathways of the WIAT intervention programme, Ward-Thompson, et al., 2019, p. 5)

Implications for Nature-Based Health and Wellbeing Intervention

In summary, a black box of controlled inputs and measured outputs is not well suited to the complexity of variables in context and process associated with nature-based and woodland activities. Nature-on-referral schemes are often run most successfully by local community groups that understand the context and needs of participants not medics (Bloomfield, 2017). As Pretty et al. (2017, p. 3) comment, 'Individual choices [...] are shaped by the interactions between the design of lived environments, transport systems, institutional inertia, advertising and corporate self-interest, and access to green space'. Our chapters on different forms of wellbeing have clearly demonstrated that processes, people and purposes are crucial in mediating experiences and their perceived benefits. What then can we conclude about whether we should engineer or engender nature-based solutions to health and wellbeing?

Abundant evidence exists for the benefits of spending time in nature and research is beginning to provide additional nuance about pathways and dosage to achieve certain outcomes. Taken together, current evidence indicates a mixed approach may be most successful. Universal access improvements such as biodiverse treed green space enhancement and early years and school nature-based curricula, coupled with collaboratively designed programmes targeting vulnerable and underrepresented groups to mediate their connection and engagement with nature, will help to build regular exposure to natural environments to effective and possibly self-sustaining levels for wellbeing. Bragg and Atkins (2016) advocate distinction between commissioned interventions for the vulnerable and public health initiatives for the general population to delineate the green care sector from more general attempts of improved societal wellbeing. Some interventions need to be specially engineered, for example where the purpose is highly specific such as increasing physical activity, or the needs of a target group are specific or sensitive. However, if we accept that our goal for wellbeing should include the more-than-human world and mutual benefit (see Chapter 2), changing our relationship with nature requires slow burn, organic and sustained approaches, an engendering of nature-based wellbeing. Broadly the two approaches seem to split along the distinction between natural infrastructural changes at a community level and green care for groups with specific needs.

Pretty et al. (2017, p. 7) in their Green Mind Theory acknowledge that changing our habits takes time and that it takes 'as a rule of thumb of 50 days at one hour per day, or 100 days (approximately 3 months) at half an hour per day, to produce changes in the brain and result in fixes to behaviour'. But the motivation to change and put in this time can be lacking and therefore support can have a significant role. For example, in the Natural Connections project (Waite, Passy, Gilchrist, Hunt, & Blackwell, 2016) where schools received training and mentoring by hub leaders and other organisations in using local natural environments for curricular teaching, it took a couple of years of ongoing support for school cultures to transform and changes in practice to become embedded and sustained. This was an example of engendered transformation, but a current DEFRA initiative Nature Friendly schools is shifting to a more engineered model with a standard quantity (dosage) of external support for schools to determine if this can result in sustainable practice in the longer term. The Woodland Trust case study in Chapter 9 also shows that intended outcomes may take years to emerge and be transformed by intervening experiences.

There is substantial evidence for starting early in providing contact and connection with nature as these experiences are formative for later use of natural environments. Education offers a good universal access point and can retro-reach other age groups and demographics through children transmitting their enthusiasm for more experiences in nature. To capitalise on that, family-friendly access to nature, through enhancing natural infrastructure, can help support more time being spent in nature. Given the urgency for a change in our relationship with nature, strategies need to address all ages and groups. This is where nuance in support may be needed to engineer access and activities to groups less likely to benefit.

'Is it possible to imagine substantial reductions in incidence of mental ill-health, obesity, type 2 diabetes, loneliness, physical inactivity, and cardiovascular disease through the adoption of resilient green mind habits, fostering prosocial and greener economies across whole populations?' asks Pretty et al. (2017, p. 12). A nature check for all policy making and implementation would go some way towards joining up wellbeing for the human and more-than-human world, since despite its complexity, its interconnectedness brings a kind of simplicity. As John Muir noted, 'When we try to pick out anything by itself, we find it hitched to everything else in the Universe' (1911, p. 110). We therefore need to be wary of simplistic input/output models; effects may be likened more to butterfly wing chaos theory in that not all the impacts will be tangible, observable or measurable and may emerge at different points throughout the lifespan. However, we can be reasonably confident that if nature-based activities are sown early and repeatedly, benefits are likely to unfold.

References

Astell-Burt, T., & Feng, X. (2019). Association of urban green space with mental health and general health among adults in Australia. *JAMA Network Open, 2*(7), e198209–e198209. Retrieved from https://jamanetwork.com/journals/jamanetworkopen/fullarticle/2739050?resultClick=3.

Barton, J., & Pretty, J. (2010). What is the best dose of nature and green exercise for improving mental health? A multi-study analysis. *Environmental Science and Technology, 44*(10), 3947–3955.

Bloomfield, D. (2017). What makes nature-based interventions for mental health successful? *British Journal of Psychiatry International, 14*(4), 82–85.

Bragg, R., & Atkins, G. (2016). *A review of nature-based interventions for mental health care* (Natural England Commissioned Reports, No. 204). Retrieved from http://publications.naturalengland.org.uk/file/6567580331409408.

Bragg, R., Wood, C., Barton, J., & Pretty, J. (2014). *Wellbeing benefits from natural environments rich in wildlife: Literature Review.*

Brindley, P., Jorgensen, A., & Maheswaran, R. (2018). Domestic gardens and self-reported health: A national population study. *International Journal of Health Geography, 17*(31). https://doi.org/10.1186/s12942-018-0148-6.

Davis, B., Rea, T., & Waite, S. (2006). The special nature of the outdoors: Its contribution to the education of children aged 3–11. *Australian Journal of Outdoor Education, 10*(2), 3–12.

Fiennes, C., Oliver, E., Dickson, K., Escobar, D., Romans, A., & Oliver, S. (2015). *The existing evidence-base about the effectiveness of outdoor learning.* London: UCL Institute of Education. Retrieved from https://givingevidence.files.wordpress.com/2015/03/outdoor-learning-giving-evidencerevised-final-report-nov-2015-etc-v21.pdf.

Frumkin, H., Bratman, G. N., Breslow, S. J., Cochran, B., Kahn, P. H., Jr., Lawler, J. J., … Wood, S. A. (2017). Nature contact and human health: A research agenda. *Environmental Health Perspectives, 125*(7), 075001.

Gibson, E., Ramsden, N., Tomlinson, R., & Jones, C. (2017). Woodland Wellbeing: A pilot for people with dementia. *Working with Older People, 21*(3), 178–185. https://doi.org/10.1108/WWOP-05-2017-0012.

Goodenough, A., Waite, S., & Bartlett, J. (2015). Families in the forest: Guilt trips, bonding moments and potential springboards. *Annals of Leisure Research, 18*(3), 377–396. http://www.tandfonline.com/doi/full/10.1080/11745398.2015.1059769.

Hillsdon, M., Panter, J., Foster, C., & Jones, A. (2006). The relationship between access and quality of urban green space with population physical activity. *Public Health, 120,* 1127–1132.

Howes, S., Edwards-Jones, A., & Waite, S. (2018a). *Moor Health and Wellbeing final report.* Plymouth: University of Plymouth.

Howes, S., Edwards-Jones, A., & Waite, S. (2018b). *Moor Health and Wellbeing toolkit.* Plymouth: University of Plymouth.

Hughes, J., Rogerson, M., Barton, J., & Bragg, R. (2019). Age and connection to nature: When is engagement critical? *Frontiers in Ecological Environments, 17*(5), 265–269. https://doi.org/10.1002/fee.2035.

Husk, K., Lovell, R., Cooper, C., & Garside R. (2013). Participation in environmental enhancement and conservation activities for health and well-being in adults. *Cochrane Database Systematic Review,* 2. Retrieved from https://www.cochrane.org/CD010351/PUBHLTH_participation-environmental-enhancement-and-conservation-activities-health-and-well-being-adults.

Jorgensen, A., & Gobster, P. H. (2010). Shades of green: Measuring the ecology of urban green space in the context of human health and well-being. *Nature and Culture, 5*(3), 338–363.

Kuo, F. E. (2003). The role of arboriculture in a healthy social ecology. *Journal of Arboriculture, 29*(3), 148–155.

Lachowycz, K., & Jones, A. P. (2013). Towards a better understanding of the relationship between greenspace and health: Development of a theoretical framework. *Landscape and Urban Planning, 118,* 62–69.

Lin, B. B., Gaston, K. J., Fuller, R. A., Wu, D., Bush, R., & Shanahan, D. F. (2017). How green is your garden? Urban form and socio-demographic factors influence yard vegetation, visitation, and ecosystem service benefits. *Landscape and Urban Planning, 157,* 239–246.

Lumber, R., Richardson, M., & Sheffield, D. (2017). Beyond knowing nature: Contact, emotion, compassion, meaning, and beauty are pathways to nature connection. *PLoS One, 12*(5), e0177186. Retrieved from https://doi.org/10.1371/journal.pone.0177186.

Marmot, M. (2010). *Fair society, healthy lives: Strategic review of health inequalities in England post-2010.* Retrieved from www.instituteofhealthequity.org.

MENE. (2018). *Monitoring Engagement in the Natural Environment.* Retrieved from https://assets.publishing.service.gov.uk/government/uploads/system/uploads/attachment_data/file/738891/Monitorof_Engagementwiththe_Natural_Environment_Headline_Report_March_2016to_February_2018.pdf.

Mitchell, R., & Popham, F. (2007). Green space, urbanity and health: Relationships in England. *Journal of Epidemiology and Community Health, 61,* 681–683.

Moeller, C., King, N., Burr, V., Gibbs, G. R., & Gomershall, T. (2018). Nature-based interventions in institutional and organisational settings: A scoping review. *International Journal of Environmental Health Research, 28*(3), 293–305.

Morgan, A., & Waite, S. (2017). Nestling into the lifeworld: The importance of place and mutuality in the early years. In V. Huggins & D. Evans (Eds.), *Early childhood education and care for sustainability* (Tactyc Series). Abingdon: Routledge.

Muir, J. (1911). *My first summer in the Sierra.* Retrieved from http://vault.sierraclub.org/john_muir_exhibit/writings/my_first_summer_in_the_sierra/.

POSTnote 0538 (Parliamentary Office for Science and Technology). (2016). *Green space and health.* Retrieved from https://researchbriefings.parliament.uk/ResearchBriefing/Summary/POST-PN-0538.

Pretty, J., Rogerson, M., & Barton, J. (2017). Green mind theory: How brain-body-behaviour links into natural and social environments for healthy habits. *International Journal of Environmental Research and Public Health, 14,* 706. https://doi.org/10.3390/ijerph14070706.

Public Health England (PHE). (2017). *Prevention concordat for mental health.* Retrieved from https://www.gov.uk/government/collections/prevention-concordat-for-better-mental-health.

Richardson, J., Goss, Z., Pratt, A., Sharman, J., & Tighe, M. (2013). Building HIA approaches into strategies for green space use: An example from Plymouth's (UK) Stepping Stones to Nature project. *Health Promotion International, 28*(4), 502–511.

Richardson, M., & Mcewan, K. (2018). 30 days wild and the relationships between engagement with nature's beauty, nature connectedness and well-being. *Frontiers in Psychology, 9,* 1500. https://doi.org/10.3389/fpsyg.2018.01500.
Silveirinha de Oliveira, E., Aspinall, P., Briggs, A., Cummins, S., Leyland, A. H., Mitchell, R., … Thompson, C. W. (2013). How effective is the Forestry Commission Scotland's woodland improvement programme—'Woods In and Around Towns' (WIAT)—At improving psychological well-being in deprived urban communities? A quasi-experimental study. *British Medical Journal Open, 2013*(3), e003648. https://doi.org/10.1136/bmjopen-2013-003648.
The Wildlife Trust. (n.d.). Retrieved from https://www.wildlifetrusts.org/.
Tillmann, S., Tobin, D., Avison, W., & Gilliland, J. (2018). Mental health benefits of interactions with nature in children and teenagers: A systematic review. *Journal of Epidemiological Community Health, 72,* 958–966.
Typhina, E. (2017). Urban park design+ love for nature: Interventions for visitor experiences and social networking. *Environmental Education Research, 23*(8), 1169–1181. https://doi.org/10.1080/13504622.2016.1214863.
Valuing Nature Network. (2017). *The cost-effectiveness of addressing public health priorities through improved access to the natural outdoors* (Report No. 6). Retrieved from http://valuing-nature.net/sites/default/files/documents/Reports/Naturally_healthy/EKN%20Naturally-Healthy-Report-Final-96dpi.pdf.
Vardakoulias, O. (2013). *The economic benefits of ecominds: A case study approach.* NEF Consulting. Retrieved from https://www.mind.org.uk/media/338566/The-Economic-Benefits-of-Ecominds-report.pdf.
Waite, S., Passy, R., Gilchrist, M., Hunt, A., & Blackwell, I. (2016). *Natural Connections Demonstration Project 2012–2016: Final report* (Natural England Commissioned Report NECR215). http://publications.naturalengland.org.uk/publication/6636651036540928.
Walton, S. (2017). *Building wellbeing together.* Retrieved from https://culturenaturewellbeing.wordpress.com/2017/09/27/building-wellbeing-together/.
Ward-Thompson, C., Roe, J., & Aspinall, P. (2013). Woodland improvements in deprived urban communities: What impact do they have on people's activities and quality of life? *Landscape and Urban Planning, 118,* 79–89.
Ward-Thompson, C., Silveirinha de Oliveira, E., Tilley, S., Elizalde, A., Botha, W., Briggs, A., … Mitchell, R. (2019). Health impacts of environmental and social interventions designed to increase deprived communities' access

to urban woodlands: A mixed methods study. *NIHR Journals: Public Health Research, 7*(2). Retrieved from https://doi.org/10.3310/phr07020.
Wells, N. M., & Lekies, K. S. (2006). Nature and the life course: Pathways from childhood nature experiences to adult environmentalism. *Children, Youth and Environments, 16*(1), 1–24. Retrieved from http://www.jstor.org/stable/10.7721/chilyoutenvi.16.1.0001.
White, M. P., Alcock, I., Grellier, J., Wheeler, B. W., Hartig, T., Warber, S., … Fleming, L. E. (2019). Spending at least 120 minutes a week in nature is associated with good health and wellbeing. *Scientific Reports, 9*, 7730. Retrieved from https://doi.org/10.1038/s41598-019-44097-3.
Zhang, G., Poulsen, D. V., Lygum, V. L., Corazon, S. S., Gramkow, M. C., & Stigsdotter, U. K. (2017). Health-promoting nature access for people with mobility impairments: A systematic review. *International Journal of Environmental Research and Public Health, 14,* 703. https://doi.org/10.3390/ijerph14070703.

11

Implications for Woodland Wellbeing Practice and Policy

In this concluding chapter, we reintroduce the Good from Woods framework of woodland wellbeing and ways in which the case study data helped to confirm and develop aspects of our indicators (feelings and experiences that support a dimension of wellbeing). We also introduce some summaries of who, what and where appeared to lead to different dimensions of woodland wellbeing. We consider some implications of GfW findings for woodland activity and health promotion. Finally, we invite those seeking to commission, provide or access woodland wellbeing to continue developing measures for and practices towards sustainable wellbeing.

Towards an Understanding of Woodland Wellbeing

Practitioner-researchers used the Good from Woods research template to focus their exploration of health and happiness service provision in a woodland context. It scaffolded attention to expectations and purposes (why), the people involved (who), the forest settings (where), the content of sessions (what) and the pedagogies and approaches used (how) (Waite, Bølling, & Bentsen, 2016). It enabled structured analysis of the human and

A. Goodenough and S. Waite, *Wellbeing from Woodland*,
https://doi.org/10.1007/978-3-030-32629-6_11

more-than-human influences within each service provision microculture that contributed towards woodland wellbeing. It supported practitioner-researchers from their position within service delivery to collect data and explore it, testing the Good from Woods working definitions of woodland wellbeing. As described in Chapter 4, practitioner-researchers' negotiation of a path between practice and research brought both challenges and opportunities.

Sometimes practitioner-researchers found it challenging to categorise who, what and where contributed to wellbeing, as well as the types of woodland wellbeing being achieved using the pilot wellbeing indicators and categories. For instance, was 'feeling in the moment', identified as important within the Embercombe case studies (Chapter 5), a sign of wellbeing or a step towards it? Or both? Was some people's expression of nature relatedness indications of their social rather than biophilic wellbeing, when nature felt like part of the intervention's collective social network? Sometimes different categories of wellbeing seemed interlinked within wellbeing experiences, frequently acting in association. Does, for example, feeling creative in natural settings provide a route to psychological wellbeing or does biophilic wellbeing underpin feeling creative in nature or does one process sometimes entangle with the other and what is the direction of flow?

At other points, temporal aspects of wellbeing seemed complicated. Sometimes, it was good to feel 'in the moment' in the woods (case study 2 at Embercombe, Chapter 5). Yet, at other times memories of the past- and future-orientated imaginings played an important role in securing positive feelings (case study 1 at Embercombe where memories were recalled in the woods and Woodland Trust and Folly Wood case studies, where woodland activity was appreciated in the context of future benefits, Chapters 5, 9, and 10). Elsewhere, feelings threatening wellbeing, evident in the moment, were later understood as part of a journey towards feeling good (such as cold and discomfort at Ruskin mill, Chapters 6). Similarly, indications of one type of wellbeing later appeared repositioned as another (such as the play that was initially associated with positive emotions at Otterhead, Chapter 8, and was later linked to nature relatedness).

Despite the challenges, however, practitioner-researchers found using the woodland wellbeing framework an important tool in understanding and defining woodland wellbeing, and in turn, they grounded and refined it with their local social and environmental knowledge.

> GfW wellbeing indicators and categories were important for exploring the data - adding my own categories and indicators from emerging themes was important. Being responsive to what emerged [R2].

Appreciating that wellbeing domains and indicators might sometimes be entangled and associated helped to ease the use of the framework. Positive emotional feelings and moods, for instance, were fairly frequently associated with other domains of health and happiness as anticipated within Fredrickson's broaden and build model (explored in Chapter 3, where positive emotions aid a flexibility of thought and action supporting people to experience other forms of wellbeing, Fredrickson & Joiner, 2002). Biophilic wellbeing, with its relaxing and restorative effects, can similarly be seen with the data to underpin increased capacity to access other forms of wellbeing in the woods. In such instances, different forms of wellbeing might be interlinked in sequential steps, one towards another, and circular flows, one reinforcing another. During 'Into the Woods' activities with the Forest of Avon Trust, 'connecting through shared experience' (an indicator created by the practitioner-researchers there) was understood to be a collective interpersonal encounter with associated emotional and social feelings. Practitioner-researchers found that young people, sharing an activity, experienced positive moods and feelings together which then supported formation of a communal bond. In turn, they experienced further social wellbeing and positive moods when practising and enjoying this collective connection.

A broad conception of what activity in woodland is and how people may engage with it also helped practitioner-researchers search widely for indications of wellbeing or disbenefits and map who, what, where and how wellbeing benefits took place during woodland experiences. For example, activity might range from fleeting, less conscious and indirect experiences (snapping a stick, for instance) to consciously focused, sustained and repeated behaviours (such as becoming a proficient coppicer). Engagement could take place through senses, body, emotions, imagination and intellect. Engagement with nature, for instance, might apply in any and all of the woodland sites in which the research took place but manifested in varied ways (from embodied physical activity at Otterhead, sensory enjoyment of natural beauty at Folly Wood to future-orientated imaginings in Woodland Trust evidence, Chapters 7, 9 and 10). Profundity of impact was not always linked to length/depth of exposure to or engagement with environment, people, practices or ethos. Tree planting with the Woodland Trust for example was a relatively short, one-off intervention that had profound attitudinal consequences. Tree planters were significantly more likely to view tree planting as part of a solution to climate change, a crisis impacting their wellbeing negatively.

In fact, contradictions and ambiguity within the evidence could sometimes lead to greater appreciation of subtle relations between nature and culture in natural health interventions. CCANW case studies findings, at first sight, appeared inconsistent. Forest Football was a disruption of everyday behaviour that allowed people freedom to enjoy increased physical activity in the forest. However, following dancers through the forest's trails, arguably a similar subversion, seemed to frustrate enjoyment of increased physical activity. Comparing the data helped reveal that the unusual activity still allowed forest footballers to respond to enough 'natural' cues to satisfy nature relatedness. Those following Dancing Trails by contrast sometimes felt cut off from such cues and an anticipated connection to nature. Taken together, these findings provide information that could be used to more closely align interventions with intended outcomes and address unintended side effects, and we have tried throughout the book to draw out such key messages from each case study.

For evaluation around access to and use of natural health services to be useful, it is also important to establish not only for whom it works but also those for whom it is less effective. If interventions feel a poor fit with participants' identity, their need to make sense in the world and their ongoing narrative of progress through it, then people may not view it as relevant or likely to be effective (case study 1 Embercombe, Courage Copse case study, Chapters 5 and 7) (Bauer, McAdams, & Pals, 2008; Ryan & Deci, 2004). These mismatches and disjunctions signal a lack of cultural congruence challenging participants' adoption of new ways of feeling or behaving (Waite, 2015). Rather than being a novel and culturally light context with few expectations and established norms of behaviour

(Waite, 2013), an intervention might demand practices that clash significantly with how participants would normally behave, and this can cause discomfort and resistance (Waite, 2015). In the majority of GfW case studies, wooded environments provided a culturally light environment (one with relatively few associations, where people could escape reminders of everyday behaviours and anxieties), but this was not always the case. Girls at secondary schools local to Courage Copse, for example, told the practitioner-researcher that classmates might only sign up to the project if they perceived themselves to be 'outdoorsy', but not if they wanted to be 'feminine'. The intervention worked hard to create a culture within the woods (through providing historical precedents of female forestry and contemporary role models) with which to address such preconceptions, hoping participants would thereby broaden their conception of who has a place in the woods and share those ideas back in the classroom with peers. However, participants' self-selection based on cultural congruence (ibid.) challenges the intervention's capacity to change minds through exposure to a different set of norms. Exploring potential service users' cultural expectations of woodland-based service delivery is important to ensure inclusivity, alongside attending to the possible benefits of cultural lightness or the need for sustained and repeated experiences to build up alternative cultural densities (sociocultural norms) with which to redress long-held beliefs.

Grounding Woodland Wellbeing

The original framework of indicators was based on relevant literature and initial feedback from practitioners. It was intended as a live document to be challenged and adapted in the light of the empirical evidence from Good from Woods case studies during the project. The figure below reintroduces the Good from Woods framework of woodland wellbeing updated with practitioner-researchers' findings. While the dimensions of health and happiness that might be accessed during woodland-based activity have stayed the same, the case studies have expanded upon the feelings, moods and behaviours that might indicate people are having an experience supportive of that category of wellbeing. Some of the why (purposes), who (people), what (activities) and how (approaches) (Waite, Bølling, & Bentsen, 2016) practitioner-researchers established as contributing to such impacts amongst trees, woods and forests are also summarised beneath these indicators (Table 11.1).

The project enabled the broader concepts associated with wellbeing in the literature on green health and wellbeing to be tested across numerous contexts and groups of participants and helped to refine them in relation to

Table 11.1 GfW post-project woodland wellbeing domains and indicators

Emotional wellbeing	Experiencing: positive emotions and moods, absence of negative emotions and moods, feeling even-tempered, relaxed, optimistic, in the moment via *chances to participate, play, create, work, imagine, engage, be outside, away from everyday settings/concerns, in the woods*
Social wellbeing	Feeling: confident, accepted, safe and supported within and through social relationships, socially supportive of others, connecting with others through shared experience via *activity that is enjoyable, shared, cooperative, inclusive, purposeful, interpersonal interactions that are enjoyable, encouraging, validating, caring*
Psychological wellbeing	Feeling: in control, competent (and seen by others to be competent), purposeful, developing oneself, secure with limitations, connecting with others through shared beliefs and outlook via *positive interpersonal interaction & regard, learning, meaningful, absorbing activities, increased influence over own and cooperative actions*
Physical wellbeing	Feeling: physical healthy, confident in and enjoying physical activity via *playful, creative, exploratory, purposeful, unrestricted movement*
Biophilic wellbeing	Feeling: absorbed in, refreshed, relaxed by the natural world, engaged in a relationship with nature, part of a bigger picture via *playful, physical, purposeful, imaginative, appreciative, spiritual and recalled engagement with the natural world, opportunities to learn about, nurture, act in the interests of other/all species*

woodland-based service provision. It is also an important contribution to the practice of woodland wellbeing that the indicators and domains have been developed with practitioners themselves so that the behaviours and language used are likely to be recognisable to other providers of nature-based wellbeing.

GfW Resources

For those interested in using the approaches developed through the GfW project in their own work, the GfW toolkit, which was collaboratively developed with practitioner-researchers, can be found at https://www.plymouth.ac.uk/research/peninsula-research-in-outdoor-learning/good-from-woods/the-toolkit. It includes much more detail on methodology, methods and analysis appropriate to the challenges and opportunities of researching woodland wellbeing. Articles have also been published by the research team in collaboration with practitioner-researchers that explore some of the case studies from different perspectives and include empirical evidence that it was not possible to explore in these chapters (Aronsson, Waite, & Tighe-Clarke, 2015; Goodenough, Waite, & Bartlett, 2015; Goodenough, Waite, & Wright, under review; Waite & Goodenough, 2018; Waite, Goodenough, Norris, & Puttick, 2016; Wright, Goodenough, & Waite, 2015).

Implications for Woodland Activity and Health Promotion

Sustainable Wellbeing

Good from Woods evidence confirmed the vital part that the materiality of trees, woods and forests plays in wellbeing outcomes associated with woodland-based health service provision. Treed environments provided biophilic effects, including restoration and relaxation, which were described as beneficial in themselves, but also underpinned access to experiences supportive of other types of wellbeing, as well as more nuanced biophilic impacts such as care for the environment. Woodland environments provided inspiration, cues and materials towards activities, both independent and guided by others, that enabled service users to meet emotional, social, psychological, physical and biophilic wellbeing needs. In collaboration with various arboreal environments, staff and volunteers developed and delivered woodland-based services to respond to the wellbeing needs of specific user groups. These included management of ancient native broadleaf woodland that caused young participants to wonder at the wood's diversity yet coherence (Embercombe, Chapter 5); planting of new trees in gardens, playgrounds and industrial sites that provided material for pupils' future good feelings (Woodland Trust, Chapter 9); regeneration of coppice that gave participants a reassuring sense of their fit within combined human/woodland logic (Ruskin Mill, Chapter 7); access to trees that provided an enduring partner in children's play despite their sometimes rough treatment (Fort Apache, Chapter 9); use of green infrastructure that invited nearby school children to move more frequently (WHY case study, Chapter 8) and an invitation for local people of all ages to connect with local environment and people, growing more intimate with each in a long unmanaged plantation (Folly Wood, Chapter 10).

> I mean even though the wood itself, it's not like, there's no wow factor up there. It's just a little –in essence, it's just a little wood that's in a way nothing special… it's more knowing what's going on, it's knowing what it's about that makes it special (Folly Wood member).

Trees, woodlands and forests actively contributed to aspects of service users' health and happiness not only as a setting that stimulates evolved, adaptive responses beneficial to humans, but also in their affordance of activities and relationships benefiting wellbeing. As Gagliano (2013) points out, we can but struggle to step outside of our anthropocentric viewpoints and simply experience ourselves as part of nature. However, by paying close attention to how we and others perceive and respond to the varying invitations that trees and woodland provide, we may begin to act in concert with them and understand better how these intra-actions contribute to our health and happiness. This is even more critical as trees, woodlands and forests and human symbiosis with them are under huge threat as we have described in Chapter 2. We want to reiterate here that there is no security of the natural environment contributing to human wellbeing without the health and thriving of that environment. It is becoming increasingly clear that threats to tree health posed by disease and climate change are associated with negative physiological and psychological outcomes for humans (Donovan et al., 2013, Howard, Rose, & Rivers, 2018). It is not only through loss of their provisioning of timber and food for instance or for example their regulation of clean air through carbon dioxide sequestration services that human health is compromised. In one US study (Donovan et al., 2013), it appears likely that the more culturally attached populations are to trees and therefore more likely to derive wellbeing from them, the higher likelihood of increased mortality within those groups when trees die in disease outbreaks. Threats to the health of trees, woods and forests are threats to multiple facets of human health and wellbeing. At the very least, a changing woodland landscape with loss of key species threatens the affordances, range of activities and experiences of nature relatedness and care that GfW research acknowledges are key aspects of woodland health and wellbeing service provision.

Local Social and Environmental Knowledge in Health Provision

GfW findings also demonstrate how local social and environmental knowledge are important resources towards evaluation of provision. The outcomes that stakeholders (including users) and service providers anticipated from services were often well aligned to those established by the research findings, although sometimes how these outcomes were achieved (how, who, what, where) was less familiar or expected amongst stakeholders. Validation of expectations does not mean that research efforts simply looked for evidence that confirmed what practitioner-researchers thought they already know. The initial process of surfacing assumptions amongst stakeholders (discussed in Chapter 4) helped to make preconception more transparent and open to critique. In addition, case studies regularly highlighted findings that expanded the range of wellbeing outcomes anticipated or perceived to be important, demonstrating the depth and breadth of evidence collection and critical analysis. Furthermore, practitioner-researcher, third sector and university research team co-production offered checks and balances for the research. Collaboration expanded conceptions and knowledge of what significant woodland health and happiness impacts might look, sound and feel like across this partnership and ensured questioning and critiquing of assumptions, methods and analyses, adding rigour.

The richness and insight into impacts provided by the local knowledge of stakeholders and service providers are also a key part of successful delivery. Wellbeing outcomes across the case studies had commonalities, but also variations that could be accounted for by differences in purposes, activity, ethos and environment. Understanding of how particular aims, settings, activities and styles of delivery may impact different service users has frequently been built up over time within organisations, projects and partnerships and is a valuable resource. Engagement with service users, including both formal and informal feedback, has frequently contributed to the re/design and intended outcomes of service delivery. Interventions explored by GfW were usually locally specific programmes, growing and changing in relation to particular places and people.

Points for practice: Nuancing provision for woodland wellbeing

- Wellbeing domains and indicators may sometimes be associated with each other. An experience supportive of wellbeing in one area (such as positive emotions) can support an experience of wellbeing elsewhere (psychological wellbeing for example). Awareness of these nuances is helpful in designing programmes.
- Woodland's support of biophilic wellbeing and its relaxing and refreshing effects commonly underpin increased capacity to access wellbeing in other areas. Similarly, activities, places and people may all contribute to positive emotions and moods that in turn support an increased capacity to access other forms of wellbeing. Taking note of purpose, people, place, content and process are vital to help understand and explain outcomes.
- Temporal aspects of wellbeing should also be considered. Experiences of wellbeing or disbenefits in the short-term may be recalled with different

associations in the long-term. Where possible, measure at different time intervals.

- Including ways to capture people's reactions to woodland experience and activity through the senses, body, emotions, imagination and intellect can help locate the important influences associated with wellbeing or disbenefits. Consider how data is collected and whether it taps into these different modes of engagement.
- Wellbeing may be sparked by fleeting, less conscious and indirect experiences as well as focused, sustained and repeated behaviours. The impact of an experience associated with wellbeing may not always be related to our length of exposure to it. Openness to unexpected sources of wellbeing can widen appreciation of how services are supporting wellbeing.
- Contradictions and ambiguities in evidence can reveal clues to woodland wellbeing and the cues trees, woods and forests provide for activity and behaviour. Accept the unexpected.
- It is important to evidence not only outcomes of woodland wellbeing, but barriers to its achievement. Interventions should consider possible cultural clashes and ways to address such preconceptions to provide inclusive provision.

Initiatives that were flexibly responsive to users' preferences enabled participants to develop activity towards meeting their wellbeing needs. Sometimes, as at Otterhead (Chapter 8), the exact pathways towards wellbeing outcomes of some user-led activity were not immediately discernible to service providers. However, careful research provided new insights into what was being achieved.

Support for the development and delivery of natural health services within UK public health policy is an indication that there is increasing recognition of how effective this sector can be in meeting health and wellbeing needs. There is burgeoning research establishing what amount and type of nature can best benefit participants, and how activity and person-centred variables mediate that exposure, which contributes to knowledge of what works, where and for whom. However, it is important that this refinement does not lead to attempts to simplify natural health service delivery into one-size-fits-all blueprints. From our research across different types of woodland-based practice, we would argue that the specificity and flexibility of their health and happiness services are two key strengths. Approaches embedded within local environmental and social knowledge with their capacity to respond in a nuanced fashion to specific user needs should not be lost in an attempt to measure and medicalise delivery into standardised remedies.

We can learn lessons from all forms of natural health service delivery towards establishing a suitable repertoire of interactions between people, activity and nature that can secure sustainable wellbeing, with benefits to both humans and non-humans. Woodland is an important aspect of natural health resourcing and as we have discussed, mutuality in our relationship for wellbeing must include trees themselves thriving. We hope that this book will inspire continued use and development of understandings of woodland wellbeing and encourage others to discover and further clarify what is good for/from woods.

References

Aronsson, J., Waite, S., & Tighe Clarke, M. (2015). Measuring the impact of outdoor learning on the physical activity of school age children: The use of accelerometry. *Education and Health, 33*(3). http://sheu.org.uk/x/eh333ja.pdf.

Bauer, J. J., McAdams, D. P., & Pals, J. L. (2008). Narrative identity and eudaimonic well-being. *Journal of Happiness Studies, 9*(81), 81–104.

Donovan, G. H., Butry, D. T., Michael, Y. L., Prestemon, J. P., Liebhold, A. M., Gatziolis, D., & Mao, M. Y. (2013). The relationship between trees and human health: Evidence from the spread of the emerald ash borer. *American Journal of Preventive Medicine, 44*(2), 139–145.

Fredrickson, B. L., & Joiner, T. (2002). Positive emotions trigger upward spirals toward emotional well-being. *Psychological Science, 13*(2), 172–175.

Gagliano, M. (2013). Persons as plants: Ecopsychology and the return to the dream of nature. *Landscapes: The Journal of the International Centre for Landscape and Language, 5*(2). Retrieved from http://ro.ecu.edu.au/landscapes/vol5/iss2/14.

Goodenough, A., Waite, S. & Bartlett, J. (2015). Families in the forest: Guilt trips, bonding moments and potential springboards. *Annals of Leisure Research,*

18(3), 377–396. http://www.tandfonline.com/doi/full/10.1080/11745398.2015.1059769.

Goodenough, A., Waite, S., & Wright, N. (under review). Place as partner: Material and affective intra-play between young people and trees. *Children's Geographies.*

Howard, C., Rose, C., & Rivers, N. (2018). *Lancet Countdown 2018 Report: Briefing for Canadian Policymakers.* Retrieved from http://www.lancetcountdown.org/media/1418/2018-lancet-countdown-policy-brief-canada.pdf.

Ryan, R. M., & Deci, E. L. (2004). An overview of self-determination theory: An organismic dialectical perspective. In R. M. Deci & R. M. Ryan (Eds.), *Handbook of self-determination research* (pp. 3–36). Suffolk: The University of Rochester Press.

Waite, S. (2013). 'Knowing your place in the world': How place and culture support and obstruct educational aims. *Cambridge Journal of Education, 43*(4), 413–434.

Waite, S. (2015). Culture clash and concord: supporting early learning outdoors in the UK. In H. Prince, K. Henderson, & B. Humberstone (Eds.), *International Handbook of Outdoor Studies.* London: Routledge.

Waite, S., & Goodenough, A. (2018). What is different about Forest School? *Journal of Outdoor and Environmental Education, 21*(1), 25–44. Retrieved from http://link.springer.com/article/10.1007/s42322-017-0005-2.

Waite, S., Goodenough, A., Norris, V., & Puttick, N. (2016). From little acorns: Environmental action as a source of ecological wellbeing. *Pastoral Care in Education: An International Journal of Personal, Social and Emotional Development, 3*(1), 43–61. Retrieved from http://www.tandfonline.com/doi/full/10.1080/02643944.2015.1119879.

Wright, N., Goodenough, A., & Waite, S. (2015). Gaining insights into young peoples' playful wellbeing in woodland through art-based action research. *Journal of Playwork Practice, 2*(1), 23–43. Retrieved from http://www.ingentaconnect.com/content/tpp/jpp/2015/00000002/00000001/art00003.

Index

A. Goodenough and S. Waite, *Wellbeing from Woodland*,
https://doi.org/10.1007/978-3-030-32629-6

C

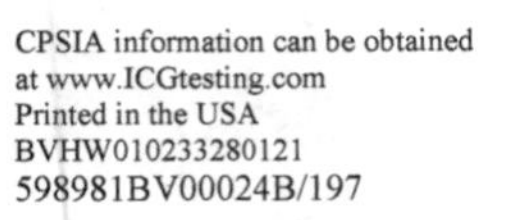

CPSIA information can be obtained
at www.ICGtesting.com
Printed in the USA
BVHW010233280121
598981BV00024B/197

9 783030 326319